Engineering Applications of Diamond

Edited by Awadesh Mallik

Published in London, United Kingdom

IntechOpen

Supporting open minds since 2005

Engineering Applications of Diamond
http://dx.doi.org/10.5772/intechopen.91098
Edited by Awadesh Mallik

Contributors
Tatsuhiko Aizawa, Tadahiko Inohara, Tomoaki Yoshino, Yohei Suzuki, Tomomi Shiratori, Funsho Olaitan Kolawole, Shola Kolade Kolawole, Luis Bernardo Varela, Marco Antonio Ramirez, Adebayo Felix Owa, André Paulo Tschiptschin, Huijie Zheng, Arne Wickenbrock, Dmitry Budker, Vadali Venkata Satya Siva Srikanth, Awadesh Mallik, Georgios Chatzidrosos, Lykourgos Bougas, Nathan Leefer, Samer Afach, Andrey Jarmola, Victor M. Acosta, Jingyan Xu, Geoffrey Z. Iwata, Till Lenz, Zhiyin Sun, Chen Zhang, Takeshi Ohshima, Hitoshi Sumiya, Kazuo Nakamura, Junichi Isoya, Jörg Wrachtrup

First published in London, United Kingdom, 2021 by IntechOpen
IntechOpen is the global imprint of INTECHOPEN LIMITED, registered in England and Wales, registration number: 11086078, 5 Princes Gate Court, London, SW7 2QJ, United Kingdom
Printed in Croatia

British Library Cataloguing-in-Publication Data
A catalogue record for this book is available from the British Library

Additional hard and PDF copies can be obtained from orders@intechopen.com

Engineering Applications of Diamond
Edited by Awadesh Mallik
p. cm.
Print ISBN 978-1-83968-531-6
Online ISBN 978-1-83968-532-3
eBook (PDF) ISBN 978-1-83968-533-0

For EU product safety concerns:
IN TECH d.o.o., Prolaz Marije Krucifikse Kozulić 3, 51000 Rijeka, Croatia,
info@intechopen.com or visit our website at intechopen.com.

Meet the editor

Dr. Awadesh Kumar Mallik is a ceramic engineer from the University of Calcutta who chose to become a diamond researcher. He started growing diamonds in the laboratory at the Indian Institute of Science Bangalore during his master thesis project for the development of vacuum tribology coatings for the Indian Space Research Organisation. He continued his journey of making diamond components for high-power electron tubes at the Council of Scientific and Industrial Research Central Glass & Ceramic Research Institute (CSIR-CGCRI), as part of a national mission for the international fusion energy International Thermonuclear Experimental Reactor (ITER) program. Dr. Mallik has also developed laser cutting and polishing technologies for as-grown diamond material over large areas for industrial collaboration. His current research focus is using diamonds for the thermal management of power electronics. He obtained a Ph.D. from Jadavpur University, India, and was an FWO postdoctoral research fellow at Hasselt University, Belgium. He has published forty peer-reviewed articles and presented many papers at international conferences.

Contents

Preface

Diamond offers many advantages over other wide-bandgap materials such as silicon carbide (SiC), gallium nitride (GaN), and others. Noticeable among these advantages is the method of fabrication. Fabrication costs of diamond-based electronics have decreased due to the advent of chemical vapour deposition (CVD). Other advantages include large Young's modulus and hardness values, which makes diamond ideal for protective coating applications. Diamond tools are used extensively in mining and construction engineering. Moreover, due to the extremely low coefficient of friction, diamond is used as a solid lubricant in outer space. Its superior mechanical properties, as well as high thermal conductivity, make diamond the best candidate for thermal management in extreme environments like outer space. Diamond is also radiation-hard; therefore, it is used as a high-energy particle detector in extreme physics experiments like fusion and fission.

The transparency of diamond across the wide electromagnetic wave spectrum makes it a material of choice for high-power laser windows, microwave power transmission gyrotron windows, and more. It is electrically insulating with high breakdown voltage and low dielectric constant. Doping diamond with boron or phosphorous makes it a useful semiconductor. Its chemical stability also makes it ideal for highly aggressive environments. The most striking property of diamond-based electrochemical sensors is diamond's high overpotential for hydrogen and oxygen evolution. There are also reports of capturing CO_2 greenhouse gas with diamond material by using the artificial photosynthesis principle. Due to its inertness in the biological environment, there have been attempts at using diamonds in the fabrication of neurons and other biological sensors. Nanocrystalline diamond's field electron emission properties allow the material to be used in cold cathode tube engineering applications. It has the highest sound velocity and thus it is the material of choice for surface acoustic wave (SAW) devices.

In recent years, much effort has been made in exploiting diamond's different vacancy defect centers for quantum applications. This book examines the properties of diamond and its numerous potential applications.

Awadesh Mallik
IMO-IMOMEC,
Hasselt University,
Diepenbeek, Belgium

Section 1

Introduction

Chapter 1

Introductory Chapter: Engineering Applications of Diamond

Awadesh Kumar Mallik

1. Introduction

Science is the knowledge of the existing laws and principles, whereas, engineering is the application of such scientific knowledge in building/designing/creating something useful for the humans, and such engineered tools/devices/processes are collectively known as technology. Knowledge about diamond materials is in existence since its early discovery along the river beds or in the mines, as early as from the 4th century BC (**Figure 1**). It was the hardest known stone which were "artfully" cut and polished to shine so beautifully that lured the kings and queens over centuries [1]. The cutting and polishing technology of these rarely found stones was popularly used to make jewellery. The diamond dust particles that are generated during jewellery stone making or the small sized stones from mines which can not be used in jewellery making, are always used as abrasives. Because of its extreme hardness, it was used for engraving other stones or grinding other materials. However, the industrial use of diamond in cutting tools has become possible with the advent of high-pressure high-temperature (HPHT) diamonds in the late 20th Century [2]. The rarity of mined diamonds made them precious and was unaffordable for the average income people. A new process called chemical vapour deposition (CVD) [3] has made it possible now to grow gem quality diamonds for the affordable jewellery application [4]. There is another method where oxygen deficient TNT/RDX

Figure 1.
A brief timeline with respect to the different milestones in diamond material history.

explosive is detonated to create diamond nanoparticles (**Figure 2**). These detonation nanodiamonds (DND) [5] are now extensively used as nucleation seeds for the CVD growth of diamond. However, the detonation process (neither ultrasonic cavitation [6] nor microplasma processing [7]) can not make gemstone quality (**Figure 3**) larger diamond crystals, whereas, HPHT can make gemstones, but they are limited in size [8] and of inferior quality diamond with defects or foreign inclusion [9]. Beauty lies in the eyes of the beholder. For the millennium generation, diamond jewellery is losing its charm and attraction. Other than aesthetic value, because of the stone's other exceptional material properties [10], like thermal conductivity, optical transparency over wide electromagnetic spectrum, velocity of sound waves, tensile strength, doping conductivity etc., diamond can also be used for the greater

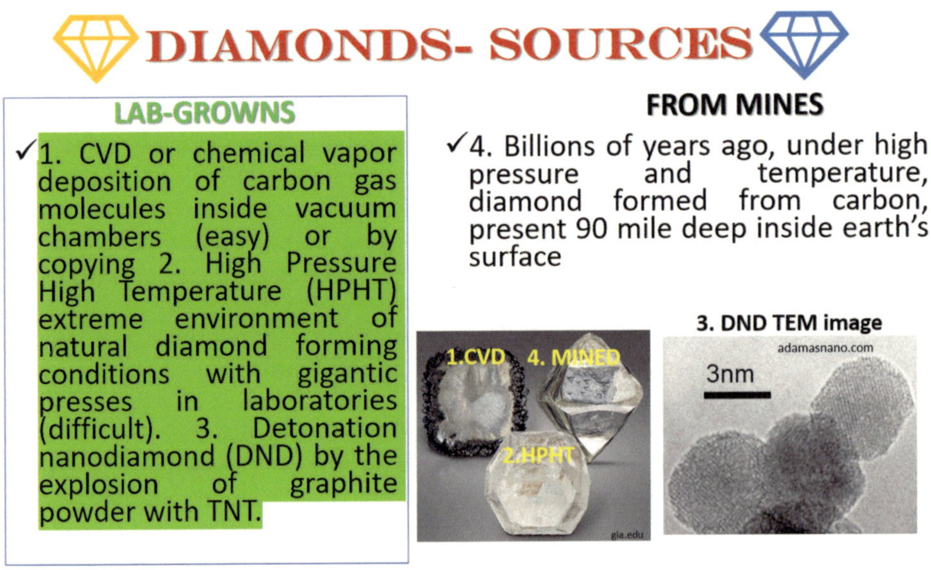

Figure 2.
Four different sources of diamond. (images are from GIA and Adamas websites).

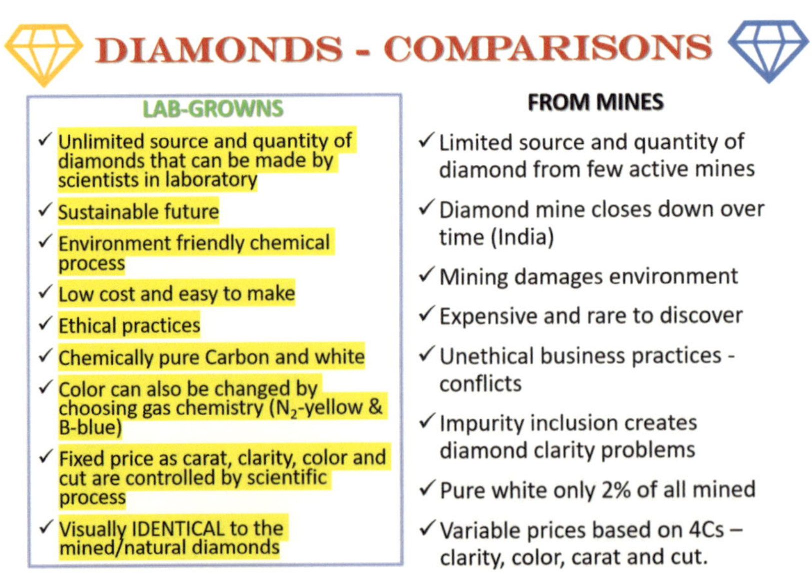

Figure 3.
A relative evaluation of the laboratory grown with their mined source of the material.

benefits of the human society [11, 12], like making faster and smaller future electronics, quantum computers, high power lasers, nuclear energy, capturing carbon for reducing its footprint in the environment [13], medical devices for patients [14] or even water purification [15] for a better standard of living (**Table 1**). This chapter lists some of the engineering applications where the scientific knowledge of the diamond material property has been used to build/design/create something useful for the people on earth.

Property	**Value**	**Application**
Hardness* [16]	100 GPa	Grinding abrasive, Cutting tool [17, 18], Tribology [19–21], mechanical applications [22]
Young's Modulus*	1100 GPa	
Poisson's ratio	0.1	
Co-efficients of friction	0.1	
Wear resistance*	10^{-7} mm^3/N-m	
Thermal conductivity at 300 K*	2000 W/m-K	Heat spreader [23], High temperature application [24, 25]
Thermal expansion co-efficient at 300 K	0.8×10^{-6} /K	
Specific heat at 20 °C	0.502 J/g-K	
Debye temperature*	1860 ± 10 K	Acoustic devices [26, 27]
Sound velocity*	17,500 m/s	
Density	3.515 g/cm^3	
Atomic Density*	1.77×10^{23} cm^{-3}	
Bandgap	5.45 eV	Power electronics packaging [28–30]
Electrical resistivity	10^{13}–10^{16} Ω-cm	
Breakdown Voltage*	10^7 V/m	
Doped [31–33] semiconductor resistivity	10^{-1}-10^4 Ω-m	Electronic sensors, devices [34–36]
Negative electron affinity	−1.5 eV (H-terminated)	Electron field emitter [37]
IR to UV optical transparency	UV cut off @ 225 nm & absorptions at 2.5–6.5 μm with theoretical 71% transmission	Photonics [38, 39], Power transmission windows [40], Jewellery [41], Quantum computing [42–44]
Absorption co-efficient	≤ 0.10 cm^{-1} at 10 μm	
Refractive index	2.38 @ 10 μm, 2.41 @ 500 nm	
Photoluminescence	Nitrogen NV, silicon SiV vacancy centres	
Corrosion resistant	Chemically inert to acids	Electrodes in electro-chemical cells [45]
Biocompatible	Inert to biological cells [46]	Medical devices [14]
Radiation hard	43 eV atomic displacement energy	Nuclear detector [47, 48], instruments [49], Betavoltaics power supply [50]
Nuclear battery	Encapsulation of radio-isotopes	
Extreme conditions	Graphitisation at T > 700 °C in an oxygen containing, and 1500 °C in an inert atmosphere	High pressure cell anvils [51, 52]

Table 1.
Diamond property and their engineering applications (highest among all materials).*

2. Mechanical engineering

The oldest (engineering) application of diamond has been cutting and polishing. Diamond is the hardest and the strongest materials with highly covalent C-C bonding. It is strong along certain crystallographic planes, in certain directions, due to variable packing density of carbon atoms. Present day scientific knowledge about the diamond crystal structure, chemistry and its other material property was developed much later, than the art of making diamond jewellery was mastered by the ancient craftsmen since the middle ages. Geologists developed the Mohs scale of hardness as shown in **Figure 4** on the basis of the relative hardness between different minerals.

However, much later on when the modern-day science started to develop, scientist found that the indentation hardness values in GPa [53] is the highest for diamond materials. **Figure 5** compares the GPa hardness of different engineering materials. It can be found that hardened steel has only 7 GPa of hardness whereas, diamond has as high as 115 GPa. However, depending on the various factors like, the amount of defects present inside like dislocations, foreign elements, single or polycrystalline diamond, CVD or HPHT grown, crystallographic planes and directions, the hardness values can vary from 25–100 GPa. Due to such high hardness value, it has been used as grinding, lapping and polishing material in the form of slurries, paste, impregnated metallic disc or paper as shown in **Figure 6**.

Diamond carbon atoms are arranged in two inter-penetrating FCC crystal lattice of a diamond cubic structure where the covalent bond length is 0.154 nm with tetrahedral angle of 109.5° between them. The highly covalent nature (deep and symmetric potential well) of the C-C bond makes diamond's Young's modulus tensile strength and the thermal expansion co-efficient, the highest among all the solid materials. The high molecular weight polyethylene polymer which are used for protective armour application has the least tensile strength (about 1 GPa) in the **Figure** 7. The woods (11 GPa) that are used to build houses, or the human teeth enamel (55 GPa) for breaking food and even the structural steel material (200 GPa) have much less stiffness or flexibility i.e. the ability to resist deformation than diamond (>1200 GPa).

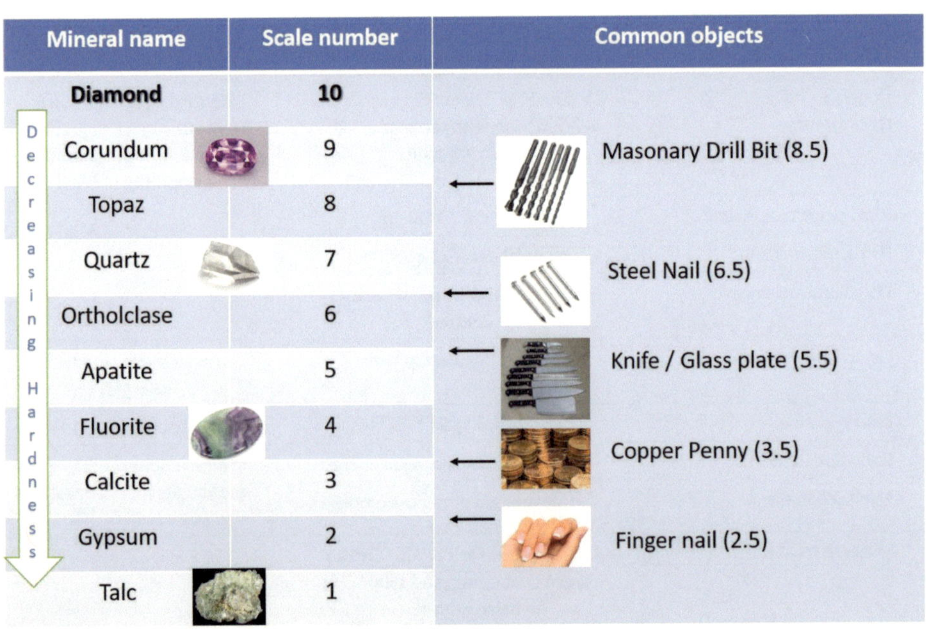

Figure 4.
Mohs scale of hardness of different materials.

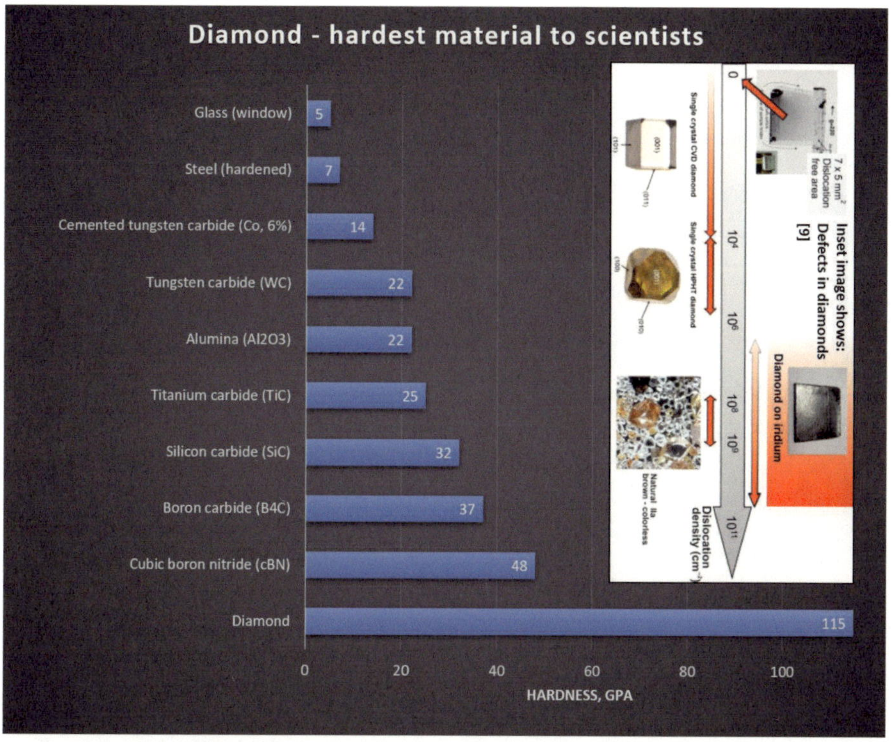

Figure 5.
A relative comparison of the hardness of different materials.

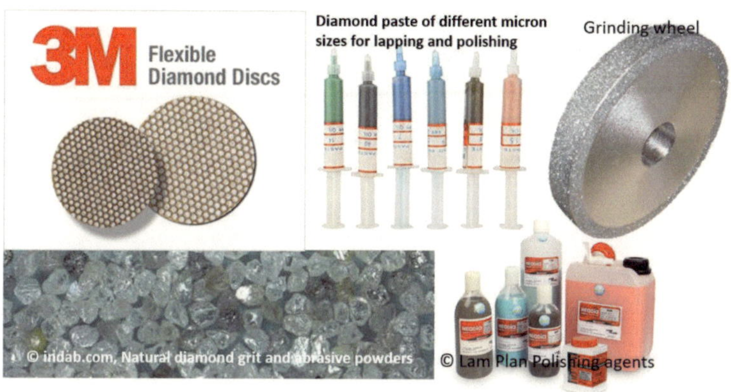

Figure 6.
Diamond abrasive application.

Due to its extreme hardness and strength, it has been traditionally used as cutting tools [54] in machining application as shown in **Figure 8**. Polycrystalline diamond cutting tools of different shapes and sizes are shown. They are used for the processing of natural stones starting from the block extraction in quarries through the intermediate steps of production to the final step of polishing the final product. Diamond tools are extensively used in the construction industry for the cutting and drilling of the concretes, asphalt and other materials. The traditional use of diamond has been for polishing glass, ceramics and the other hard metals, as already described before. Various types of metal bonded or pre-alloyed (cobalt) powders are mixed with synthetic diamond powder by hot pressing or sintering for

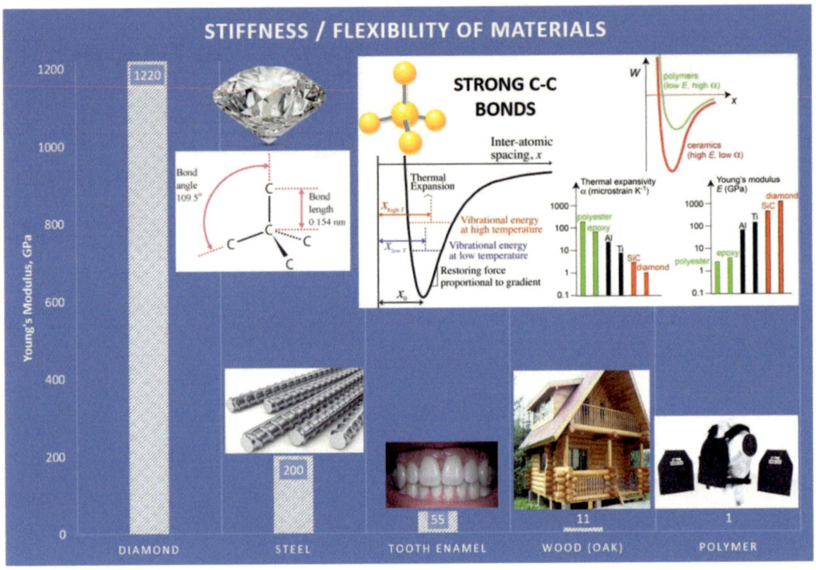

Figure 7.
A relative comparison of the strength of different materials.

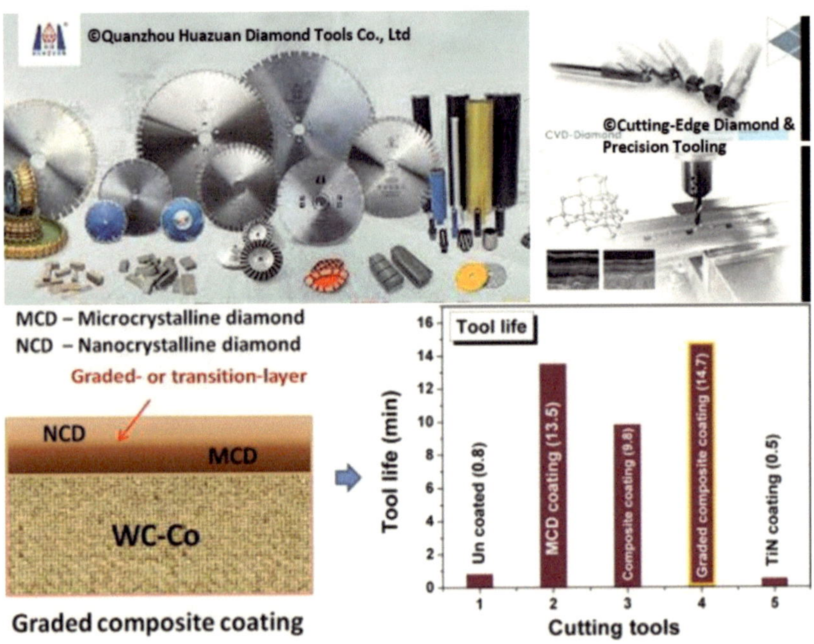

As reported by, R. Dumpala et al., International Journal of Refractory Metals and Hard Materials, 48, 2015, 24-30, https://doi.org/10.1016/j.ijrmhm.2014.07.023.

Figure 8.
Diamond cutting tools application.

the abrasive industry. It has been shown by the researchers (**Figure 8**) that a double-layer diamond coating with micro (MCD) and nanocrystalline diamond (NCD) grains on the top of traditional Co cemented WC cutting tools not only increases the tool life but also it enhances the cutting efficiency. Such coated tools can be recycled time and again after recoating with diamond, once the top coating is worn out. Diamond is the best protective solution for the coating service industry. Wear is the major cause of economic loss due to the energy that is lost in overcoming the mechanical friction within the moving mechanical assemblies. Diamond tribology [55, 56] is an important engineering application.

3. Electrical engineering

When current passes through electronic circuits, it heats up the devices, which even sometimes lead up to the device failures. Future generation devices will be smaller and faster, therefore there will be more current passing through per unit area of electronic circuits that will heat up the devices enormously. For efficient working of our devices, this heat needs to be thrown out of the electronic circuits, and diamond does this job the best, being the material with highest thermal conductivity (**Figure 9**). Moore's law earlier predicted that every 2 years the size of the electronics will be reduced by half. Diamond can only keep the pace of the Moore's law with time. Direct contact of diamond with electronic chips will pass the heat away from the circuits and thereby making the current to flow easily within your device for smooth operations (**Figure 10**). There are commercial suppliers, like Element Six, of such CVD diamond heat spreaders. If we compare the cooling capacity of different coolants, it is observed that diamond is five times more effective than commonly used Cu in electrical engineering. Polycrystalline diamond is alloyed/mixed with Cu/Ag/Ti metal powders and then sintered together for making composites for electronic packaging application [57, 58]. Therefore, engineers are designing future technologies like 5G/7G, radars for space or military communication with the integration of diamond in their electronic circuits. The scanning electron micrograph in **Figure 9** shows one such CVD grown polycrystalline diamond plate like microstructure suitable for heat spreading applications.

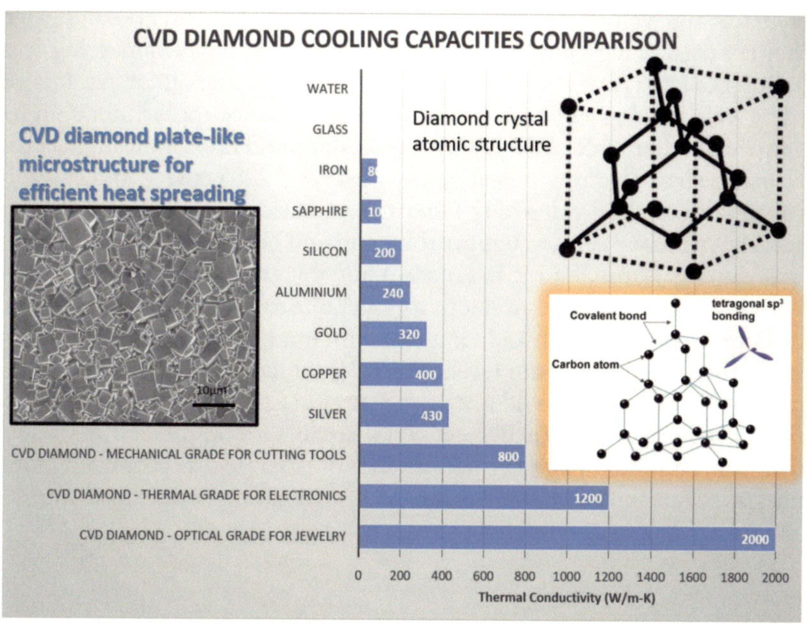

Figure 9.
A relative comparison of the thermal conductivity of different materials.

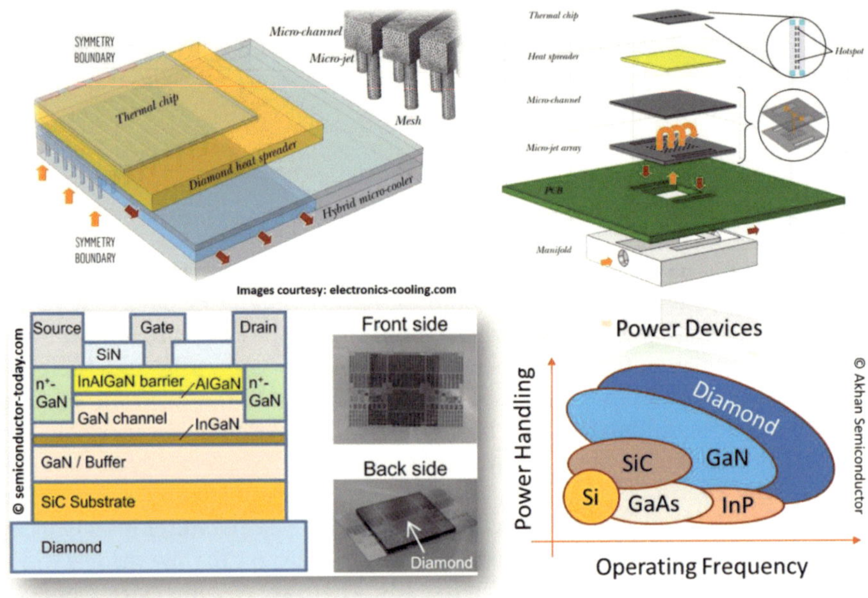

Figure 10.
Diamond thermal management application.

4. Energy/power engineering

As we know from our high school physics that the energy level difference between the conduction band and valence band divides materials into a. insulator - with large differences, b. semiconductor - with small differences and c. metal - with overlapping of the bands. Materials like GaN (3.44 eV), SiC (2.36 eV) and Diamond (5.45 eV) have large values of band gaps and they are known as wide band gap materials. They are used in high power high temp. high freq. energy engineering applications. In order to keep pace with the Moore's law, Si is running out of gas. It is getting replaced by wide band gap materials for high power density applications (**Figure 10**). Compare to other wide band gap materials, diamond is with the highest band gap, also has the best electron–hole mobility (1945 and 2285 cm / V. s), critical breakdown voltage and the best value of the thermal expansion co-efficient. However, it is intrinsically insulator at room temperature and will become semiconductor only by suitable doping. Boron doping has made it possible to produce acceptor levels suitable for room temperature conductivity (metallic to superconductor [59, 60] based on doping concentration and temperature); but phosphorous doped n-type diamond has deep electron donor levels (0.46 eV) which only become active at high temperatures. Nitrogen also could not dope diamond to produce n-type conductivity, rather it produces NV centre defects [38] - suitable for opto-electronic engineering or quantum computer engineering. Absence of suitable n-type dopant atom for diamond, has so far limited the future prospect of diamond based electronic devices. It can only be used as single electrode - but not as transistors.

5. Computer engineering

In a maze puzzle, in order to find out "the only way out of the confinement", one has to explore all the different possible routes, one at a time, in order to look for the "single" viable solution - which is time consuming. Classical computing would take long time to find a solution by trial and error, on the basis of its binary states of "0" and "1". Quantum mechanics gives wave particle duality i.e., quantum entities or qubits can be present simultaneously at more than one location, therefore, if qubit tries to find the way out of a maze puzzle, due to its entanglement and superposition characteristic of different states at the same time, it will be possible to find/compute the solution of maze puzzle much faster and in efficient manner. In other words, if an electron is asked to find the way out of a maze, due to its quantum nature, it will visit all the routes inside the maze simultaneously and will return with the correct maze path solution within no time! Quantum computing based on qubit has many advantages over classical computing. It can process much bigger amount of data at much less amount of time. In today's world of artificial intelligence and machine learning with increasing amount of data, the classical computing is reaching its limit of computational power. Therefore, there is greater need of increasing the computational power of today's computers. And the solution lies in quantum computers. The search for qubits started in 1980s and there are trapped ions, quantum dots or cryogenic superconductor-based quantum information processing, however, diamond advantageously offers a nitrogen vacancy NV centre based solid state room temperature qubit [61, 62]. First ever continuous-wave (CW) room-temperature solid-state maser using the NV defect in diamond was reported in 2018 [63]. There are numerous large and small start-up companies, supported by national and international government agencies [64], who are devoting research effort in coming up with a viable diamond-based quantum computer in the coming decade or so.

6. Chemical engineering

Boron doped diamond electrodes are used for many electrochemistry-based applications [65] like sensing, environmental, electrosynthesis, electrocatalysis for energy and devices. Chemo-mechanical polishing [66] by diamond slurries uses the combined effect of chemical reaction in addition to the mechanical abrasion of hard surfaces for polishing application.

7. Sonic/acoustic engineering

IDT metallic lines are patterned onto SAW devices. Sound velocity divided by the IDT internal spacing gives the frequency of such devices, which can be used as pressure and temperature sensors under extreme heat and pressure conditions of internal combustion engine for auto-mobile industry. The frequency of SAW devices can be enhanced by the use of diamond substrate material [67] with high sound velocity. Diamond being the material with the highest acoustic wave velocity is ideal for different sonic applications, like tweeter domes [68].

8. Opto-electronics engineering

Nitrogen vacancy (NV) centre [69] defect inside diamond crystal lattice has room temperature quantum spin states which interacts with presence of an external magnetic field. Higher the external magnetic field higher is their interaction. The energy which is required to flip the NV centre spin state would also become higher. This energy of interaction can be probed by electron paramagnetic resonance spectroscopy (EPR) – when input microwave energy/frequency ($E = h\nu$) matches with the interaction energy, input microwave energy flips then NV centre spin state and thereby the intensity of fluorescence drops which is detected by optical microscope. Thus, the resonance frequency provides a direct and quantitative measurement of the local external magnetic field. NV centre magnetometer has been so far explored for jam-less GPS navigation by interacting with the earth's magnetic field (Lockheed Martin is developing), surface scanning probes to magnetically characterise semiconductors, oxides and other materials, spintronics, nanoscale thermometry, marker for living cells etc.

HPHT or CVD diamond optical lenses [70] are used for wide range of spectrum from infrared to UV windows for their unique optical properties, chemical, mechanical and thermal stability under extreme conditions of high-power optical beams. They can be used as visible intraocular lens, X-ray refractive lens [71] and even for spectrometers.

9. Bioengineering

Artificial retina based on silicon chips was earlier coated with ultra nanocrystalline diamond (UNCD) for eye environment fluid protective application [72]. Nowadays diamond electrode [73] is even tried for electrical stimulation of retinal prosthetic implants [74]. Diamond surfaces have been functionalised [75] for various applications [76, 77] like biomarker, bio-chip using electrochemical reactions. Microwave plasma CVD grown single crystal diamonds [78] is also used as dosimeter detector in radiotherapy treatments for cancer [79].

10. Environmental engineering

Diamond coatings have been developed by many companies to treat industrial waste water and also to disinfect freshwater without use of any chemicals. The boron doped diamond electrodes oxidise the organic pollutant into CO_2 or destroys the dirt and disinfect the germs that are present in the water. Recently a European project titled "DIACAT" has used the same boron doped diamond for direct photocatalytic conversion of CO_2 into fine chemicals and fuels under visible light [80].

11. Nuclear engineering

Diamond has the best mechanical properties alongwith high thermal conductivity [81] and very low dielectric loss tangent [82], which make them the only material that can be used as power transmission windows in the gyrotrons used for fusion reactors [25]. Synthetic diamond detector designed for small field dosimetry is used as a dosimeter for synchrotron microbeam and minibeam radiotherapy to ensure highly localised and precise dose delivery [83]. Betavoltaics are converting the beta particle (high energy electrons) decay of the radioactive material into the electric current of a semiconductor material (electron–hole pair generation by ionisation), that lasts for the half-life time period of the radioactive material itself. Researchers at Bristol, UK, [84] have separated C14 radio-isotope from the nuclear power plant waste material to form diamond out of them, which can be used as a nuclear battery to power low-capacity device application for space, military or medical, like hearing aid in human body for their entire lifetime. But safety is still the main concern for its actual use.

These are some (**Figure 11**), among the many, engineering applications of diamond that are available and/or under testing, for better technologies of the future.

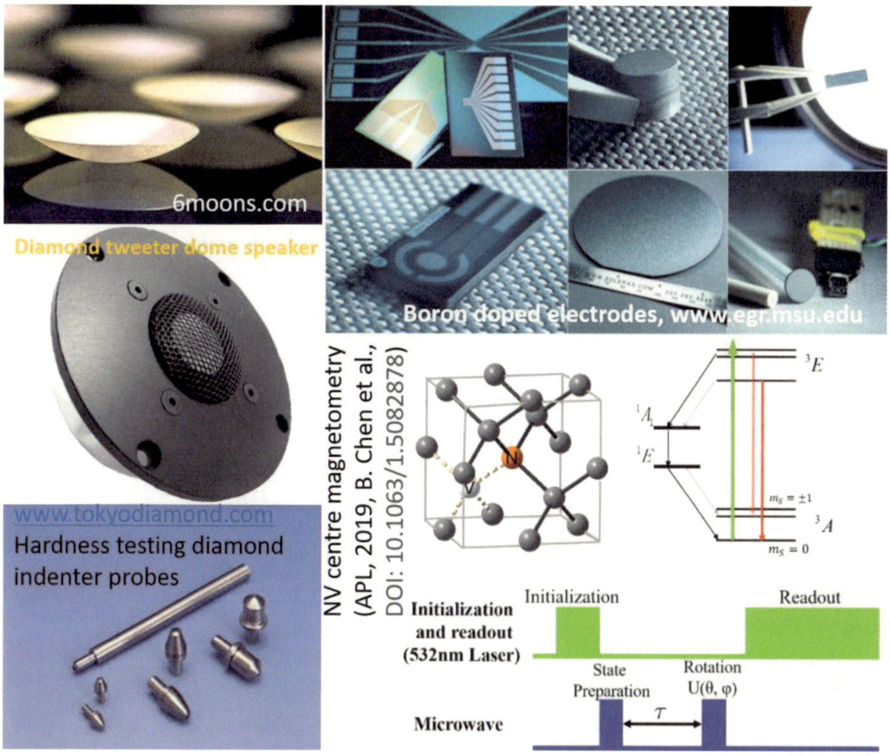

Figure 11.
Few important engineering applications of diamond.

Acknowledgements

AKM thankfully acknowledge the Research Foundation Flanders - (FWO) for his Postdoctoral Researcher fellowship grant no (12X2919N) at Hasselt University, under the supervision of Prof. Ken Haenen.

Author details

Awadesh Kumar Mallik[1,2]*

1 Hasselt University, Institute for Materials Research (IMO), Diepenbeek, Belgium

2 IMEC vzw, IMOMEC, Diepenbeek, Belgium

*Address all correspondence to: awadesh.mallik@uhasselt.be

References

[1] The history of diamond cutting and polishing technology, GIA Knowledge Sessions Webinar Series, https://www.youtube.com/watch?v=t_bOARK9DAs, Last accessed on 11th February, 2021

[2] A. I. Prikhna , High_Pressure Apparatuses in Production of Synthetic Diamonds (Review), ISSN 1063_4576, Journal of Superhard Materials, 2008, Vol. 30, No. 1, pp. 1-15. © Allerton Press, Inc., 2008.

[3] J.C. Angus, Diamond synthesis by chemical vapor deposition: The early years, Diamond & Related Materials, 49 (2014) 77-86. doi:10.1016/j.diamond.2014.08.004

[4] Laboratory-grown diamonds: Updates and Identification, GIA Knowledge Sessions Webinar Series, https://www.youtube.com/watch?v=35ZSy_y5Umc, Last accessed on 11th February, 2021.

[5] V. N. Mochalin, O. Shenderova, D. Ho, Y. Gogotsi, The properties and applications of nanodiamonds, Nature Nanotechnology 7 (2012) 11-23. doi:10.1038/nnano.2011.209

[6] A. K. Khachatryan, S. G. Aloyan, P. W. May, R. Sargsyan, V. A. Khachatryan, V. S. Baghdasaryan, Graphite-to-diamond transformation induced by ultrasonic cavitation, Diamond and Related Materials, 17 (2008) 931. doi:10.1016/j.diamond.2008.01.112

[7] A. Kumar, P. Ann Lin, A. Xue, et al. Formation of nanodiamonds at near-ambient conditions via microplasma dissociation of ethanol vapour. Nat Commun 4, 2618 (2013). https://doi.org/10.1038/ncomms3618

[8] Hideaki Yamada, Akiyoshi Chayahara, Yoshiaki Mokuno, Nobuteru Tsubouchi, Shin-ichi Shikata, Uniform growth and repeatable fabrication of inch-sized wafers of a single-crystal diamond, Diamond & Related Materials 33 (2013) 27-31

[9] Diamond wafer technologies for semiconductor device applications, Editor(s): Satoshi Koizumi, Hitoshi Umezawa, Julien Pernot, Mariko Suzuki, In Woodhead Publishing Series in Electronic and Optical Materials, Power Electronics Device Applications of Diamond Semiconductors, Woodhead Publishing, 2018, Pages 1-97, https://doi.org/10.1016/B978-0-08-102183-5.00001-7.

[10] T. Eisenberg, E. Schreiner, Book: Diamonds: Properties, synthesis and applications, 2011.

[11] R. S. Sussmann, J. R. Brandon, S. E. Coe, C. S. J. Pickles, C. G. Sweeney, A. Wasenczuk, C. J. H. Wort, C. N. Dodge, CVD Diamond: A new engineering material for thermal, dielectric and optical applications, Industrial Diamond Review, 58 (1998) 69-77.

[12] R. J. Nemanich, J. A. Carlisle, A. Hirata, K. Haenen, CVD diamond—Research, applications, and challenges, MRS Bulletin, 39 (2014) 490-494. doi: http://dx.doi.org/10.1557/mrs.2014.97

[13] Peter Knittel, Franziska Buchner, Emina Hadzifejzovic, Christian Giese, Patricia Quellmalz, Robert Seidel, Tristan Petit, Boyan Iliev, Thomas J. S. Schubert, Christoph E. Nebel, John S. Foord, Nanostructured Boron Doped Diamond Electrodes with Increased Reactivity for Solar-Driven CO2 Reduction in Room Temperature Ionic Liquids. ChemCatChem, 12, 2020, 5548-557. https://doi.org/10.1002/cctc.202000938

[14] Roger J Narayan, Ryan D. Boehm, Anirudha V. Sumant, Medical applications of diamond particles & surfaces, Materials Today, Volume 14, Issue 4, 2011, Pages 154-163, https://doi.org/10.1016/S1369-7021(11)70087-6.

[15] Martínez-Huitle, Carlos Alberto, Conductive diamond electrodes for water purification, Materials Research, 10, 2007, pp.419-424. https://dx.doi.org/10.1590/S1516-14392007000400016

[16] Alexander Quandt, Igor Popov, David Tománek, Superior hardness and stiffness of diamond nanoparticles, Carbon, 162, 2020, 497-501, https://doi.org/10.1016/j.carbon.2020.02.076.

[17] Guangxian Li, Mohammad Zulafif Rahim, Wencheng Pan, Cuie Wen, Songlin Ding, The manufacturing and the application of polycrystalline diamond tools – A comprehensive review, Journal of Manufacturing Processes, 56, 2020, 400-416, https://doi.org/10.1016/j.jmapro.2020.05.010.

[18] D. A. Lucca, M. J. Klopfstein, O. Riemer, Ultra-Precision Machining: Cutting With Diamond Tools, ASME. *J. Manuf. Sci. Eng*. November 2020; 142(11): 110817. https://doi.org/10.1115/1.4048194

[19] J. Fineberg, Diamonds are forever — or are they?, Nature Mater 10, 3-4 (2011). https://doi.org/10.1038/nmat2926

[20] Hua Wang, Xin Song, Xinchang Wang, Fanghong Sun, Tribological performance and wear mechanism of smooth ultrananocrystalline diamond films, Journal of Materials Processing Technology, 290, 2021, 116993, https://doi.org/10.1016/j.jmatprotec.2020.116993.

[21] T. Bergs, U. Müller, F. Vits, S. Barth, Tribological conditions in grinding of polycrystalline diamond, Diamond and Related Materials, 108, 2020, 107930, https://doi.org/10.1016/j.diamond.2020.107930.

[22] T. Schuelke, T. A. Grotjohn, Diamond polishing, Diamond and Related Materials, 32 (2013) 17-26. doi:10.1016/j.diamond.2012.11.007

[23] K. Jagannadham, Multilayer diamond heat spreaders for electronic power devices, Solid-State Electronics, 42[12] (1998) 2199-2208. doi:10.1016/S0038-1101(98)00216-0

[24] D. Francis, F. Faili, D. Babić, F. Ejeckam, A Nurmikko, H. Maris, Formation and characterization of 4-inch GaN-on-diamond substrates, Diamond and Related Materials, 19[2-3] (2010) 229-233. doi:10.1016/j.diamond.2009.08.017

[25] A. K. Mallik, N. Dandapat, S. Chakraborty, J. Ghosh, M. Unnikrishnan, V. K. Balla, Characterisations of microwave plasma CVD grown polycrystalline diamond (PCD) coatings for advanced technological applications, Journal of the Processing and Application of Ceramics, 8[2] (2014) 69-80. doi: 10.2298/PAC1402069M

[26] Debarati Mukherjee, Filipe J. Oliveira, Rui F. Silva, José F. Carreira, Luis Rino, Maria R. Correia, Shlomo Z. Rotter, Luis N. Alves , Joana C. Mendes Diamond-SAW devices: a reverse fabrication method, Volume13, Issue1, Special Issue: E-MRS 2015 Spring Meeting – Symposium D,January 2016, Pages 53-58, https://doi.org/10.1002/pssc.201510313

[27] A.K.Mallik, S. Roy, V. K.Balla, S.Bysakh, R.Bhar, Characteristics of CVD grown diamond films on langasite substrates, Journal of Coating Science and Technology, ISSN (online): 2369-3355, 2020, 41-51. (Impact factor: not available; Citations:0) DOI: https://doi.org/10.6000/2369-3355.2019.06.02.2[

[28] Chris J.H. Wort, Richard S. Balmer, Diamond as an electronic material, Materials Today, 11, 2008, 22-28, https://doi.org/10.1016/S1369-7021(07)70349-8.

[29] Etienne Gheeraert, Book: Power electronic devices performances based on diamond properties, January 2018,

[30] S. S. Zuo, M. K. Yaran, T. A. Grotjohn, D. K. Reinhard, J. Asmussen, Investigation of diamond deposition

uniformity and quality for freestanding film and substrate applications, Diamond and Related Materials, 17[3] (2008) 300-305. doi:10.1016/j.diamond.2007.12.069

[31] R. Kalish, Doping of diamond, Carbon 37 (1999) 781-785.

[32] Kevin G. Crawford, Isha Maini, David A. Macdonald, David A.J. Moran, Surface transfer doping of diamond: A review, Progress in Surface Science xxx (2021) xxx.

[33] Satoshi Koizumi, Tokuyuki Teraji, Hisao Kanda , Phosphorus-doped chemical vapor deposition of diamond , Diamond and Related Materials 9 (2000) 935-940.

[34] Aneeta Jaggernauth, Joana C. Mendes, Rui F. Silva, Atomic layer deposition of high-κ layers on polycrystalline diamond for MOS devices: a review, J. Mater. Chem. C, 2020,8, 13127-13153.

[35] A. V. Sumant, O. Auciello, R. W. Carpick, S. Srinivasan, J. E. Butler, Ultrananocrystalline and nanocrystalline diamond thin films for MEMS/NEMS applications, MRS Bulletin, 35 (2010) 281-288. http://dx.doi.org/10.1557/mrs2010.550

[36] Kerem Bray, Hiromitsu Kato, Rodolfo Previdi, Russell Sandstrom, Kumaravelu Ganesan, Masahiko Ogura, Toshiharu Makino, Satoshi Yamasaki, Andrew P. Magyar, Milos Toth, Igor Aharonovich , Single crystal diamond membranes for nanoelectronics, **Nanoscale, 10 (2018)** 4028-4035, https://doi.org/10.1039/C7NR09097H

[37] I. Lin, S. Koizumi, J. Yater, F. Koeck, Diamond electron emission. *MRS Bulletin, 39* (2014) 533-541. doi:10.1557/mrs.2014.101

[38] I. Aharonovich, A. Greentree, S. Prawer, Diamond photonics. Nature Photon, 5, 397-405 (2011). https://doi.org/10.1038/nphoton.2011.54

[39] Jonathan C. Lee, Andrew P. Magyar, David O. Bracher, Igor Aharonovich, Evelyn L. Hu, Fabrication of thin diamond membranes for photonic applications, Diamond & Related Materials 33 (2013) 45-48.

[40] M. Thumm, MPACVD-diamond windows for high-power and long-pulse millimeter wave transmission, Diamond and Related Materials, 10[9-10] (2001) 1692-1699. doi:10.1016/S0925-9635(01)00397-1

[41] Q. Liang, C. S. Yan, Y. Meng, J. Lai, S. Krasnicki, H. K. Mao, R. J. Hemley, Recent advances in high-growth rate single-crystal CVD diamond. Diamond & Related Materials 2009; 18: 698-703. http://dx.doi.org/10.1016/j.diamond.2008.12.002

[42] J. Wolters, M. Strau, R. S. Schoenfeld, O. Benson, Quantum zeno phenomenon on a single solid state spin, Physical Review A, 88 (2013) 020101(R). http://dx.doi.org/10.1103/PhysRevA.88.020101

[43] M.L. Markham, J.M. Dodson, G.A. Scarsbrook, D.J. Twitchen, G. Balasubramanian, F. Jelezko, J. Wrachtrup, CVD diamond for spintronics, Diamond & Related Materials 20 (2011) 134-139.

[44] B. J. M. Hausmann, M. Khan, Y. Zhang, T. M. Babinec, K. Martinick, M. McCutcheon, P. R. Hemmer, M. Loncar, Fabrication of diamond nanowires for quantum information processing applications, Diamond and Related Materials, 19[5-6] (2010) 621-629. doi:10.1016/j.diamond.2010.01.011

[45] Freitas Jhonys Machado, Oliveira Thiago da Costa, Munoz Rodrigo Alejandro Abarza, Richter Eduardo Mathias, Boron Doped Diamond Electrodes in Flow-Based Systems, Frontiers in Chemistry, 7 (2019) 190. DOI=10.3389/fchem.2019.00190

[46] Human osteoblast like MG63 cell and mouse fibroblast NIH3T3 cell viability study on the nucleation side of CVD grown polycrystalline diamond coatings, Anuradha Jana, Nandadulal Dandapat, Somoshree Sengupta, Vamsi Krishna Balla, Rajnarayan Saha, Awadesh Kumar Mallik*, Trends in Biomaterials and Artificial Organs, 2015(3) 211-216. http://www.biomaterials.org.in/ojs/index.php/tibao/article/view/617

[47] A. Oh, Particle detection with CVD diamond, Univ Hamburg Germany, Inst Experim Physics, PhD thesis, 1999. http://cds.cern.ch/record/403633/files/cer-000330326.pdf

[48] Kai Su, Qi He1, Jinfeng Zhang1,2, Zeyang Ren1,2, Linyue Liu3, Jincheng Zhang1,2, Xiaoping Ouyang3 and Yue Hao, 2021 *J. Phys. D: Appl. Phys.* 54 145105, https://doi.org/10.1088/1361-6463/abd5e2

[49] S.N. Polyakov, V.N. Denisov, N.V.Kuzmin, M.S. Kuznetsov, S.Yu. Martyushov, S.A. Nosukhin, S.A. Terentiev, V.D. Blank, Characterization of top-quality type IIa synthetic diamonds for new X-ray optics, Diamond & Related Materials 20 (2011) 726-728.

[50] V.S. Bormashova, S.Yu. Troschieva, S.A. Tarelkina, A.P. Volkova, D.V. Teteruka, A.V. Golovanova, M.S. Kuznetsova, N.V. Kornilova, S.A. Terentieva, V.D. Blanka, High power density nuclear battery prototype based on diamond Schottky Diodes, Diamond & Related Materials 84 (2018) 41-47.

[51] Simone Anzellini, Silvia Boccato, A Practical Review of the Laser-Heated Diamond Anvil Cell for University Laboratories and Synchrotron Applications, Crystals 2020, 10, 459; doi:10.3390/cryst10060459

[52] Guoyin Shen, Yanbin Wang, High-pressure Apparatus Integrated with Synchrotron Radiation, Reviews in Mineralogy & Geochemistry Vol. 78 pp. 745-777, 2014.

[53] Snigdha Roy, Vamsi K. Balla, Awadesh K. Mallik, Victor G. Ralchenko, Andrey P. Bolshakov and Eugene E. Ashkinazi, Polishing of Black and White CVD Grown Polycrystalline Diamond Coatings, Journal of Coating Science and Technology, 2018, 5, 50-58. DOI: https://doi.org/10.6000/2369-3355.2018.05.02.3

[54] J.S. Konstanty, 19 - Applications of powder metallurgy to cutting tools, Editor(s): Isaac Chang, Yuyuan Zhao, In Woodhead Publishing Series in Metals and Surface Engineering, Advances in Powder Metallurgy, Woodhead Publishing, 2013, Pages 555-585, ISBN 9780857094209, https://doi.org/10.1533/9780857098900.4.555.

[55] A. K. Mallik, S. K. Biswas, High Vacuum Tribology of Polycrystalline Diamond Coatings, Sadhana, Vol. 34, Part 5, Oct 2009, pp 811-821. doi: 10.1007/s12046-009-0047-4

[56] Anuradha Jana, Nandadulal Dandapat, Mitun Das, Vamsi Krishna Balla, Shirshendu Chakraborty, Rajnarayan Saha, Awadesh Kumar Mallik, Severe wear behaviour of alumina balls sliding against diamond ceramic coatings, Bulletin of Materials Science, April 2016, Volume 39, Issue 2, pp 573-586. https://link.springer.com/article/10.1007/s12034-016-1166-2.

[57] Yu-Siang Jhong, Hsiao-Ting Tseng, Su-Jien Lin, Diamond/Ag-Ti composites with high thermal conductivity and excellent thermal cycling performance fabricated by pressureless sintering, Journal of Alloys and Compounds 801 (2019) 589-595.

[58] Spark Plasma Sintering of Ti-diamond Composites, Awadesh Kumar Mallik, Mitun Das, Sumana Ghosh, Dibyendu Chakravarty, Ceramics International, (2019), https://doi.org/10.1016/j.ceramint.2019.02.204.

[59] T. Kageura, M. Hideko, I. Tsuyuzaki, et al. Single-crystalline boron-doped

diamond superconducting quantum interference devices with regrowth-induced step edge structure. Sci Rep 9, 15214 (2019). https://doi.org/10.1038/s41598-019-51596-w

[60] E. Bustarret, P. Achatz, B. Sacépé, C. Chapelier, C. Marcenat, L. Ortéga, T. Klein, Metal-to-insulator transition and superconductivity in boron-doped diamond, Philosophical Transactions of the Royal Society A: Mathematical, Physical and Engineering Sciences, 366 (2007) 267-279, https://doi.org/10.1098/rsta.2007.2151

[61] C. E. Bradley, J. Randall, M. H. Abobeih, R. C. Berrevoets, M. J. Degen, M. A. Bakker, M. Markham, D. J. Twitchen, and T. H. Taminiau, A Ten-Qubit Solid-State Spin Register with Quantum Memory up to One Minute, Phys. Rev. X 9, 031045 – Published 11 September 2019. DOI:https://doi.org/10.1103/PhysRevX.9.031045

[62] Elizabeth Gibney, Quantum physics: Flawed to perfection, Nature, Nature Feature, 2014.

[63] J. Breeze, E. Salvadori, J. Sathian, et al. Continuous-wave room-temperature diamond maser. Nature 555, 493-496 (2018). https://doi.org/10.1038/nature25970

[64] Elizabeth Gibney, Quantum gold rush: the private funding pouring into quantum start-ups, Nature, News Feature, 2019.

[65] N. Yang, S. Yu, J. V. Macpherson, Y. Einaga, H. Zhao, G. Zhao, G. M. Swain, Conductive diamond: synthesis, properties, and electrochemical applications, Chemical Society Reviews 48 (2018) 157-204. https://doi.org/10.1039/C7CS00757D

[66] Awadesh Kumar Mallik, Radhaballabh Bhar, Sandip Bysakh, An effort in planarising microwave plasma CVD grown polycrystalline diamond (PCD) coated Si wafers of 4 inch diameter, Materials Science in Semiconductor Processing 43 (2016) 1-7. http://dx.doi.org/10.1016/j.mssp.2015.11.016

[67] A. K. Mallik, S. Roy, V. K. Balla, S. Bysakh, R. Bhar, Characteristics of CVD grown diamond films on langasite substrates, Journal of Coating Science and Technology, ISSN (online): 2369-3355, 2020, 41-51. DOI: https://doi.org/10.6000/2369-3355.2019.06.02.2

[68] R. S. Balmer, J. R. Brandon, S. L. Clewes, H. K. Dhillon, J. M. Dodson, I. Friel, P. N. Inglis, T. D. Madgwick, M. L. Markham, T. P. Mollar, Chemical vapour deposition synthetic diamond: materials, technology and applications, 2009 *J. Phys.: Condens. Matter* 21 364221.

[69] Nitrogen-Vacancy Centers in Diamond: Nanoscale Sensors for Physics and Biology, Annual Review of Physical Chemistry, Vol. 65:83-105 (Volume publication date April 2014) First published online as a Review in Advance on November 21, 2013 https://doi.org/10.1146/annurev-physchem-040513-103659

[70] E. Woerner, C. Wild, W. Mueller-Sebert, P. Koidl, CVD-diamond optical lenses, Diamond and Related Materials, Volume 10, Issues 3-7, 2001, Pages 557-560, https://doi.org/10.1016/S0925-9635(00)00393-9.

[71] M. Polikarpov, V. Polikarpov, I. Snigireva, A. Snigirev, Diamond X-ray Refractive Lenses with High Acceptance, Physics Procedia, Volume 84, 2016, Pages 213-220, https://doi.org/10.1016/j.phpro.2016.11.037.

[72] Mark Peplow, Artificial retina gets diamond coating, 2005, Nature, doi:10.1038/news050328-9

[73] Y. Einaga, J. S. Foord, G. M. Swain, Diamond electrodes: diversity and maturity, MRS Bulletin 39 (6), 525-532.

[74] A. Ahmood et al., Diamond devices for high acuity prosthetic vision, Advanced Biosystems, Volume1, Issue1-2, February 2017, 1600003

[75] Jorne Raymakers, Ken Haenen, Wouter Maes, Diamond surface functionalization: from gemstone to photoelectrochemical applications, **J. Mater. Chem. C**, 2019,7, 10134-10165, https://doi.org/10.1039/C9TC03381E

[76] A. Kuwahata, T. Kitaizumi, K. Saichi, et al., Magnetometer with nitrogen-vacancy center in a bulk diamond for detecting magnetic nanoparticles in biomedical applications. Sci Rep 10, 2483 (2020). https://doi.org/10.1038/s41598-020-59064-6

[77] M. Sobaszek, K. Siuzdak, J. Ryl, R. Bogdanowicz, G. M. Swain, The electrochemical determination of isatin at nanocrystalline boron-doped diamond electrodes: Stress monitoring of animals, Sensors and Actuators B: Chemical 306, 127592.

[78] Awadesh Kumar Mallik, Microwave plasma CVD grown single crystal diamond coatings – a review, Journal of Coating Science & Technology, 2016, 3, 75-99. DOI: http://dx.doi.org/10.6000/2369-3355.2016.03.02.4

[79] F. Marsolat, D. Tromson, N. Tranchant, M. Pomorski, D. Lazaro-Ponthus, C. Bassinet, C. Huet, S. Derreumaux, M. Chea, G. Boisserie, J. Alvarez, P. Bergonzo, Diamond dosimeter for small beam stereotactic radiotherapy, Diamond and Related Materials, Volume 33, 2013, Pages 63-70, https://doi.org/10.1016/j.diamond.2013.01.003.

[80] Fang Gao, Christoph E. Nebel, Electrically Conductive Diamond Membrane for Electrochemical Separation Processes. *ACS Applied Materials & Interfaces*, 2016 Jul 20;8(28): 18640-6.

[81] A.F. Popovich, V.G. Ralchenko, V.K. Balla, A.K. Mallik, A.A. Khomich, A.P. Bolshakov, D.N. Sovyk, E.E. Ashkinazi, V.Yu. Yurov, Growth of 4″ diameter polycrystalline diamond wafers with high thermal conductivity by 915 MHz microwave plasma chemical vapour deposition, Plasma Science and Technology, 19, 035503, 2017, https://doi.org/10.1088/2058-6272/19/3/035503.

[82] Measurement of the Complex Permittivity of Polycrystalline Diamond by the Resonator Method in the Millimeter Range, M. P. Parkhomenko, D. S. Kalenov, N. A. Fedoseev, I. S. Eremin, V. G. Ral'chenko, A. P. Bol'shakov, E. E. Ashkinazi, A. F, Popovich, V. K. Balla, and A. K. Mallik, ISSN 1541-308X, Physics of Wave Phenomena, 2015, Vol.23, No 3, Pp. 1-6. DOI:10.3103/S1541308X15030012

[83] Livingstone J, Stevenson AW, Butler DJ, Häusermann D, Adam JF. Characterization of a synthetic single crystal diamond detector for dosimetry in spatially fractionated synchrotron x-ray fields. Med Phys. 2016 Jul;43(7):4283. doi: 10.1118/1.4953833. PMID: 27370143.

[84] "Diamond-age" of power generation as nuclear batteries developed, https://www.youtube.com/watch?v=b6ME88nMnYE, last accessed on 11th February, 2021.

Section 2

Opto-Electronics

Chapter 2

Novel Magnetic-Sensing Modalities with Nitrogen-Vacancy Centers in Diamond

Huijie Zheng, Arne Wickenbrock, Georgios Chatzidrosos, Lykourgos Bougas, Nathan Leefer, Samer Afach, Andrey Jarmola, Victor M. Acosta, Jingyan Xu, Geoffrey Z. Iwata, Till Lenz, Zhiyin Sun, Chen Zhang, Takeshi Ohshima, Hitoshi Sumiya, Kazuo Nakamura, Junichi Isoya, Jörg Wrachtrup and Dmitry Budker

Abstract

In modern-day quantum metrology, quantum sensors are widely employed to detect weak magnetic fields or nanoscale signals. Quantum devices, exploiting quantum coherence, are inevitably connected to physical constants and can achieve accuracy, repeatability, and precision approaching fundamental limits. As a result, these sensors have shown utility in a wide range of research domains spanning both science and technology. A rapidly emerging quantum sensing platform employs atomic-scale defects in crystals. In particular, magnetometry using nitrogen-vacancy (NV) color centers in diamond has garnered increasing interest. NV systems possess a combination of remarkable properties, optical addressability, long coherence times, and biocompatibility. Sensors based on NV centers excel in spatial resolution and magnetic sensitivity. These diamond-based sensors promise comparable combination of high spatial resolution and magnetic sensitivity without cryogenic operation. The above properties of NV magnetometers promise increasingly integrated quantum measurement technology, as a result, they have been extensively developed with various protocols and find use in numerous applications spanning materials characterization, nuclear magnetic resonance (NMR), condensed matter physics, paleomagnetism, neuroscience and living systems biology, and industrial vector magnetometry. In this chapter, NV centers are explored for magnetic sensing in a number of contexts. In general, we introduce novel regimes for magnetic-field probes with NV ensembles. Specifically, NV centers are developed for sensitive magnetometers for applications where microwaves (MWs) are prohibitively invasive and operations need to be carried out under zero ambient magnetic field. The primary goal of our discussion is to improve the utility of these NV center-based magnetometers.

Keywords: NV center, novel magnetometry

1. Introduction

1.1 Microwave-free magnetometry

The negatively-charged NV center in diamond, an atomic-scale defect consisting of a substitutional nitrogen adjacent to a vacancy in the diamond lattice, has emerged as a unique nanoscale sensor with numerous interesting applications in the past years and has been extensively explored which yields various technological breakthroughs. It can be used, based on employing electron spin resonance (ESR) techniques, to detect magnetic fields [1], temperature [2, 3], electric fields [4], strain [5], rotation [6, 7], and other fundamental physical quantity, for example, quantum geometrical phases [8]. In particular, the NV center in diamond has been proved to be a successful magnetic-field sensor by using the technique optically-detected magnetic resonance (ODMR) [9], for both with single and ensembles of NV centers [1, 10, 11]. The ODMR sensing protocols are typically realized by polarizing the NV electron spin with green light and manipulating the spin state with MW fields. An optical readout step involves either detection of NV-photoluminescence (PL) [1] or absorption of 1042 nm radiation resonant with the singlet transition [12–15]. When the applied MW fields are resonant with the splitting of the magnetic sublevels in the ground state of the NV-center, the transfer of spin populations results in an observable change in PL or absorption.

The use of strong MW fields is not desirable in certain cases that it can produce interference in the measurement environment which is detrimental to the sensing protocol and therefore limits the applications of an NV-based sensors. An example is, in the context of magnetic induction tomography, the detection of eddy currents in conductive materials [16–18], which is later demonstrated in the publication [19], where the application of MW to the diamond will be heavily affected by the presence of a conductive structure under investigation. Another example is imaging the pattern of conductive, magnetic structures [20].

There have been, in the literature, several demonstrations of MW-free, and all-optical, diamond-based magnetic measurement, which are implemented initially with single NV centers [21–23], and recently with NV ensembles [20]. These protocols have been realized by exploiting either the properties of the NV-center's PL or their decoherence properties at certain external magnetic fields. In particular, one of the main approaches exploits the properties of the NV-center PL in the strong magnetic-field regime for all-optical magnetic field mapping. The basic operating principle resides on the fact that under the application of a strong magnetic field (much stronger than any strain field within the diamond crystal), the transverse magnetic-field components induce mixing of the NV-center spin states. Note that this spin-mixing leads to modifications of the NV spin dynamics under optical illumination, yielding a reduction of the ODMR contrast, which eventually hampers ODMR-based magnetometric protocols. However, this reduction in contrast is accompanied by an overall reduction of the NV-center PL, which can be exploited in an all-optical magnetic-field sensing and imaging protocol. So far, these protocols remain either qualitative, requiring complicated setups to achieve high spatial resolution, or lack high magnetic-field sensitivities and bandwidth.

An alternative approach exploits the decoherence properties of the NV-center. The presence of high off-axis magnetic fields increases the amount of spin mixing and results in a reduction of the longitudinal spin relaxation time (T_1) [24]. This technique has been used to image weak fluctuating magnetic fields and to characterize spin noise in ferromagnetic materials. More recently, an all-optical, MW-free, spin-relaxation contrast magnetometer for the imaging of thin magnetic films, with a temporal and spatial resolution of 20 ms and 440 nm, respectively, was realized [20].

Recently, it has been demonstrated the principles of a sensitive MW-free magnetometer by exploiting the properties of the ground-state level anticrossing of the NV center in diamond. In particular, a ~102.4 mT background magnetic field causes degeneracy and mixing (anticrossing) of ground-state Zeeman sublevels, which is observable as a drop in the fluorescence signal. Changes in the external magnetic field will perturb the anti-crossing condition and, thus, result in a photoluminescence signal that can be monitored and used for magnetic-field measurement. This type of magnetometer offers technical and logistical advantages in bio-imaging and conductive-material sensing. Finally, the MW-free technique is deployed to perform eddy current detection. The method exhibits a combination of high spatial resolution and high sensitivity. As an example of application, detection of eddy-current induced magnetic field on metal samples, by implementing a cross-relaxation, microwave-free, magnetic detection scheme, is demonstrated. Compared to existing technologies, the demonstrated method exhibits higher, sub-millimeter spatial resolution.

A schematic of the experimental setup is shown in **Figure 1**. The diamond is placed within a custom-made electromagnet (**Figure 1(a)**) and can be rotated around the z-axis. The electromagnet can be rotated around the y-axis and also moved with a motorized 3D translation stage to align the magnetic field with the respect to the diamond. Therefore, with all the degrees of freedom, the position of

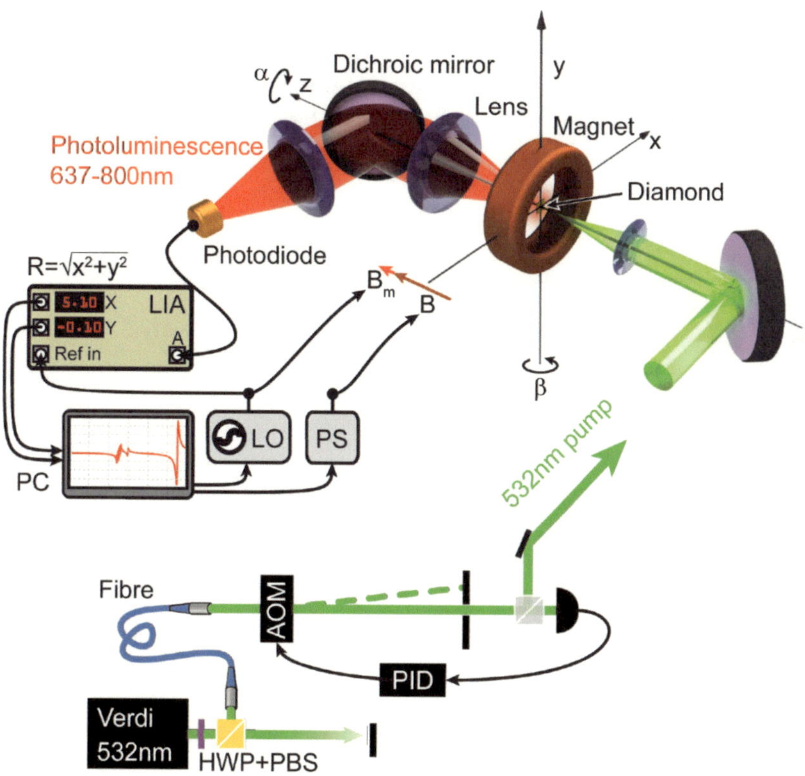

Figure 1.
Schematic of the experimental setup. The diamond is placed within a custom-made electromagnet and can be rotated around the z-axis. The electromagnet can be moved with a computer-controlled 3D translation stage to align the magnetic field with respect to the diamond. The laser light propagates through an acousto-optical modulator, part of which is monitored on a photodetector (PD). The signal is used as a monitor/measurement signal to a PID controller to stabilize the beam power. The beam is focused with a 40 mm focal-length lens into the diamond after the acousto-optic modulator (AOM). The photoluminescence is collected with a 30 mm focal-length lens (numerical aperture ~ 0.5) and focused onto a PD. The PD signal is sent to the lock-in amplifier (LIA). Figure adopted from Ref. [25].

diamond can be centralized in the magnet and the alignment of the [25] NV-axis to the magnetic field can be optimized.

To apply small modulation of the magnetic field, a secondary coil is wound around the diamond mount. An oscillating field B_m is produced by the coil which is also used as the local oscillator for the LIA.

The NV centers in diamond are optically spin-polarized with 532 nm laser light. Before the diamond, the light is sent trough an acousto-optical modulator to enable power modulation. Part of the laser light is split-off and measured on a photodiode (PD). The signal is input into a feedback controller to stabilize the beam power. After the AOM, the beam is focused with a lens into the diamond. The red/near-infrared NV-PL is collected with a 30 mm focal-length lens, which is then separated from the green transmission with a dichroic mirror and a notch-band optical filter for the green pump light before it's being focused onto a PD. The readout signal is demodulated by the LIA at the modulating frequency or measured at DC.

The magnetic field was scanned from 0 mT to more than 110 mT after initial alignment and calibration of the magnet, and the PL was monitored. It shows the measured PL measured as a function of the applied external magnetic field in **Figure 2(a)** and the corresponding LIA signal in **Figure 2(b)**. The modulation frequency of the field was 100 kHz, the modulation amplitude was about 0.1 mT.

Both plots contain several features previously discussed in the literature. The initial gradual decrease in PL is associated with a reduction in emission of the non-aligned NV centers due to spin-mixing [27]. The observed features around 51.4 mT, correspond to cross-relaxation events between the NV center and single substitutional nitrogen (P1) centers [27–29]. Several additional features are visible. They can be associated with cross-relaxation events with either the nuclear spins of nearby P1 centers [27–29] or nuclear spins of ^{13}C atoms. For details, readers can refer to Wickenbrock *et al.* [25] and Ivady *et al.* [30].

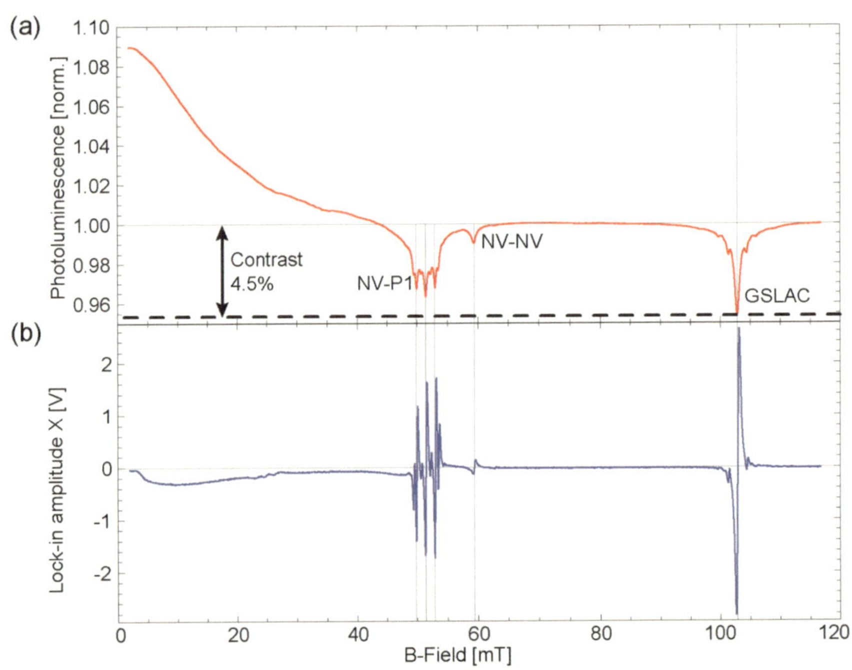

Figure 2.
(a) NV-PL as a function of the applied magnetic field normalized to the PL at 80 mT. (b) Derivative signal of (a) as given by the in-phase output (X) of the LIA. Figure adopted from Ref. [26].

Around the ground-state level anticrossing (GSLAC) feature, the derivative fluorescence signal as detected in the properly phased LIA X-output depends linearly on the magnetic field and can therefore be used for precise magnetic-field measurements. To subtract the coefficient for translating the LIA output signal into magnetic field value, the data within the near-linear region around the zero-crossing point are fitted with a straight line. Then the background magnetic field is set to the magnitude corresponds to the zero-crossing point where the magnetometer is maximally sensitive to external magnetic fields. At this field, the setup is insensitive to magnetic field variations and the data can be used to understand the technical noise level of the magnetometer. The noise floor is flat and around 6 nT/$\sqrt{\text{Hz}}$ [25] (**Figure 3**).

We now discuss an application of the presented MW-free sensing technique, that is the detection of magnetic fields generated by eddy currents in conductive materials. Eddy currents are induced in conductive objects when an oscillating or pulsed magnetic field is applied. These currents in turn produce a magnetic field in

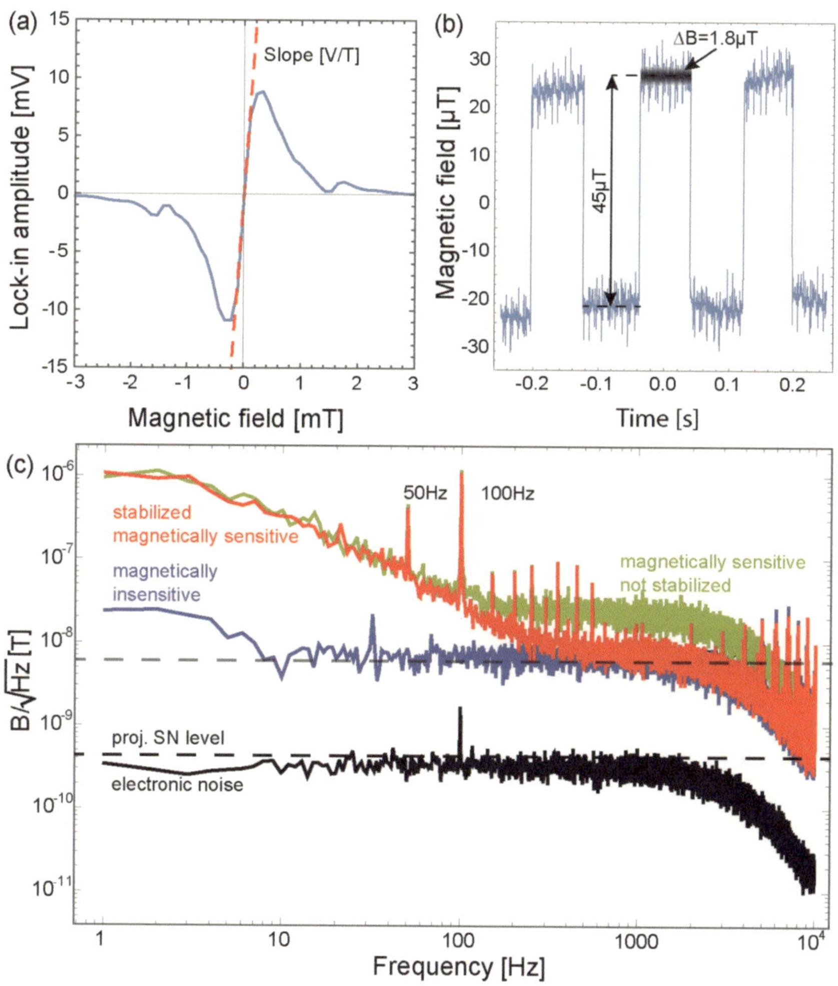

Figure 3.
Magnetometer noise characterization. (a) Calibration data and linear fitting of the GSLAC feature around 102.4 mT. (b) Magnetic sensing validation by applying a square-wave magnetic-field. (c) Noise floor of the magnetometer: Magnetically sensitive configuration with pump laser light not stabilized (green) and stabilized (red), configuration of magnetically insensitive with pump laser light stabilized (blue) and electronic noise without the pump light stabilization (black). Figures adopted from Ref. [25].

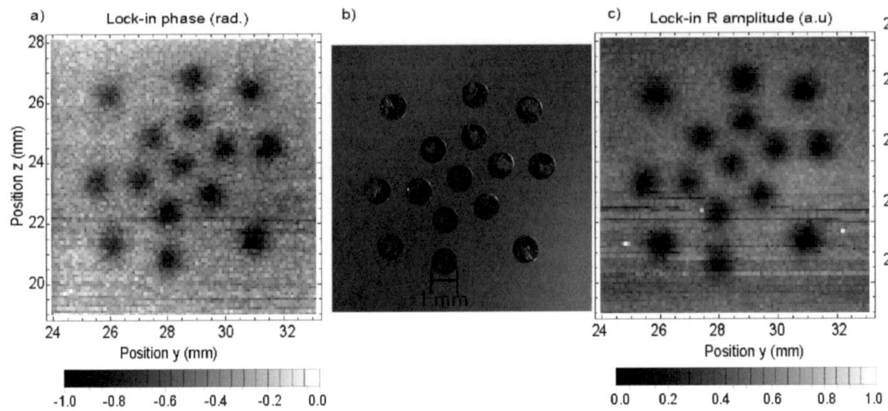

Figure 4.
(a), (c) Lock-in phase and lock-in R amplitude for a PCB containing fifteen aluminum dots imaged with eddy-current (b) photograph of the PCB with the corresponding length of 1 mm noted for scale. Figures adopted from Ref. [27].

response. A measurement of this field provides means to detect the eddy currents [18]. The induced field depends on the material's properties and shape, as well as the skin depth for the applied alternating field [19]. This has been demonstrated with vapor-cell magnetometers [18] and is a standard technique in industry with inductive coils. The detection of eddy currents with coils is an established technique. The benefits of using NV centers could be the high spatial resolution and the possibility to investigate highly conductive materials since the magnetometer can sense oscillating fields down to DC.

The eddy-current detection should be realized without MWs, since, because of the proximity to conducting materials, the amplitude and phase of the MWs are affected and therefore the magnetometric performance would be deteriorated. For this reason, the work was carried out with a MW-free NV magnetometer. As shown in **Figure 2**, the photoluminescence of NVs changes as the applied magnetic field increases. All the features in the plot, including the initial gradual decrease (at fields from zero to ≈ 25 mT), the NV-P1 feature, the NV-NV interaction feature and the GSLAC, can be utilized to probe magnetic fields. As investigated in Ref. [19], the initial gradual decrease of the NV PL is the most robust to misalignment between the applied magnetic field and the preferential NV axis among those features. Therefore, it is chosen to perform eddy current detection.

One of the advantages of NV-based sensors is their spatial resolution. For eddy-current imaging, for a constant conductivity, the smaller a material is, the smaller the amplitude of the secondary field it produces. As a test of spatial resolution, fifteen 1 mm-diameter 35 μm thick aluminum disks are imaged. The LIA amplitude and phase affected by these disks, as response of the NV-based magnetic field sensor, is shown in **Figure 4**. The overall pattern formed by the metal disks and non-magnetic materials between them, which are both on the scale of sub-millimeter, are clearly visible.

2. Microwave-free vector magnetometry

Sensitive vector magnetometers are exploited in applications including magnetic navigation [31], magnetic anomaly detection [32], current and position sensing [32], and measuring biological magnetic fields [33, 34]. Several versatile magnetometry platforms have emerged over the past decades, such as Hall probes, flux-gate, tunneling-magnetoresistance [35], SQUID based magnetometry [36] and vapor-cell-based magnetometry [37, 38]. Particularly compelling are sensors based on negatively charged NV centers in single-crystal diamond, providing high-sensitivity magnetic sensing and high-resolution imaging [15, 39, 40]. There is growing interest in magnetic-field sensors with high spatial resolution, for example to study biological processes or the composition of materials. NV magnetometry allows measuring magnetic fields at microscopic scales and/or at ambient temperature, providing new tools for probing various phenomena including semiconductor materials [41] and metallic compounds [42], magnetism in condensed matter systems [43], and elucidating spin order in magnetic materials [44].

To date, vector magnetometers based on NV diamond have been realized by interrogating ensembles of NV centers along multiple orientations [45, 46] or relying on a hybrid quantum platform involving a NV center probe and a nuclear-spin qubit at particular positions [47]. These techniques, however, are all based on ODMR technique that needs to apply microwaves simultaneously or sequentially [45, 46, 48]. The requirement of microwave control brings the possibility of spurious harmonics within the measurement and hinders applications in areas where the application of microwaves is prohibitively invasive and where it is inherently difficult to achieve such control. Although NV magnetometers have been implemented as vector magnetic probes at room temperature, it has remained challenging to achieve vector capability under cryogenic temperatures (less than 4 K) due to difficulties with thermal management. Heat dissipation from microwave wires is unavoidable and causes temperature variations, restricting the sensors for numerous innovative applications [49].

The work Zheng *et al.* [50] proposes and demonstrates a protocol that enables vectorial measurement of magnetic fields by interrogating an ensemble of NV centers aligned along only a single crystallographic axis at the GSLAC. By applying two orthogonal alternating fields, parallel and perpendicular to the chosen axis, the presented technique offers simultaneous and direct readout of full vector magnetic information, free from systematic errors during reconstruction. In contrast to existing approaches, this protocol does not employ microwave fields. Thus it is possible to extend NV-based vector magnetic sensing techniques to cryogenic temperatures.

The proposed vector magnetic sensing protocol is based on optical detection of the GSLAC of NV center in diamond. It was first proposed and demonstrated for scalar magentometry in Wickenbrock *et al.* [25]. However, both the longitudinal and transverse magnetic fields can lead to a change in the photoluminescence signal at the GSLAC. While the response to the direction of the transverse component is non-trivial [51, 52], achieving vectorial sensing of magnetic fields with a single NV center or an ensembles along a single crystallographic axis faces a number of significant challenges.

The NV center has a spin-triplet ground state (S = 1), which can be optically polarized to $|m_s = 0\rangle$ and read out owing to a spin-dependent intersystem crossing into intermediate singlet states. At zero field, the $|m_s = \pm 1\rangle$ states are (nearly) degenerate; however, these states lie higher in energy than the $|m_s = 0\rangle$ state due to spin–spin interaction. This is the zero-field splitting $\mathcal{D}_{gs}$ and corresponds to a frequency difference of 2.87 GHz. Because of Zeeman effect, a subset of NV centers' magnetic sublevels experience a complex GSLAC at an field $B_z \approx 102.4$ mT [25, 53]. The energy levels, including the coupling to the intrinsic nuclear spin of nitrogen (I=1), of the NV center are shown in **Figure 5(a)**.

Figure 5(b) shows the photoluminescence spectrum as a function of the external magnetic field. The GSLAC appears as a remarkably sharp feature around 102.4

mT, zoomed-in in the inset. In the inset, the visible satellite features can be attributed to cross-relaxation with the nearby spin bath [13, 25, 27, 28]. The $|m_s = 0\rangle$ and $|m_s = -1\rangle$ magnetic-sublevel can be coupled/mixed by transverse fields and therefore the profile of the GSLAC feature are affected which changes the contrast and amplitude. **Figure 5(c)** shows the traces of the GSLAC feature for several transverse fields in the range of ± 0.06 mT. The amplitudes of the GSLAC feature as a function of transverse field is indicated by the trace-colored dots and connected with the black line. In summary, the GSLAC contrast exhibits a relatively narrow (FWHM $\approx$ 38 μT) magnetic-resonance feature as a function of transverse magnetic field centered around zero transverse field.

To adapt a scalar magnetometer for vector measurement, it is typical to apply orthogonal fields modulated at various frequencies. Thus it is possible to determine the components along each direction by individually demodulating the signal [38]. The authors propose a method to realize vector-field sensing in the x-y plane using a transverse field rotating around the z-axis with just one frequency.

To gain an intuitive understanding, the PL lineshape is approximated as a function of transverse magnetic fields with a 2-D Lorentzian centered around $B_x = B_y = 0$, **Figure 6(a)** (i). With a transverse field applied that is rotating around z, the PL signal will be reduced but remains unmodulated, indicated by the red curve in **Figure 6(a)** (ii). In the presence of an additional static transverse field, the PL signal shows a modulation at the rotation frequency with a minimum when the rotating field points in the same direction as the field under interrogation and a maximum when both are antiparallel, shown in **Figure 6(a)** (iii), (iv) and (v). The difference between the PL signals with (red curve) and without (blue curve)

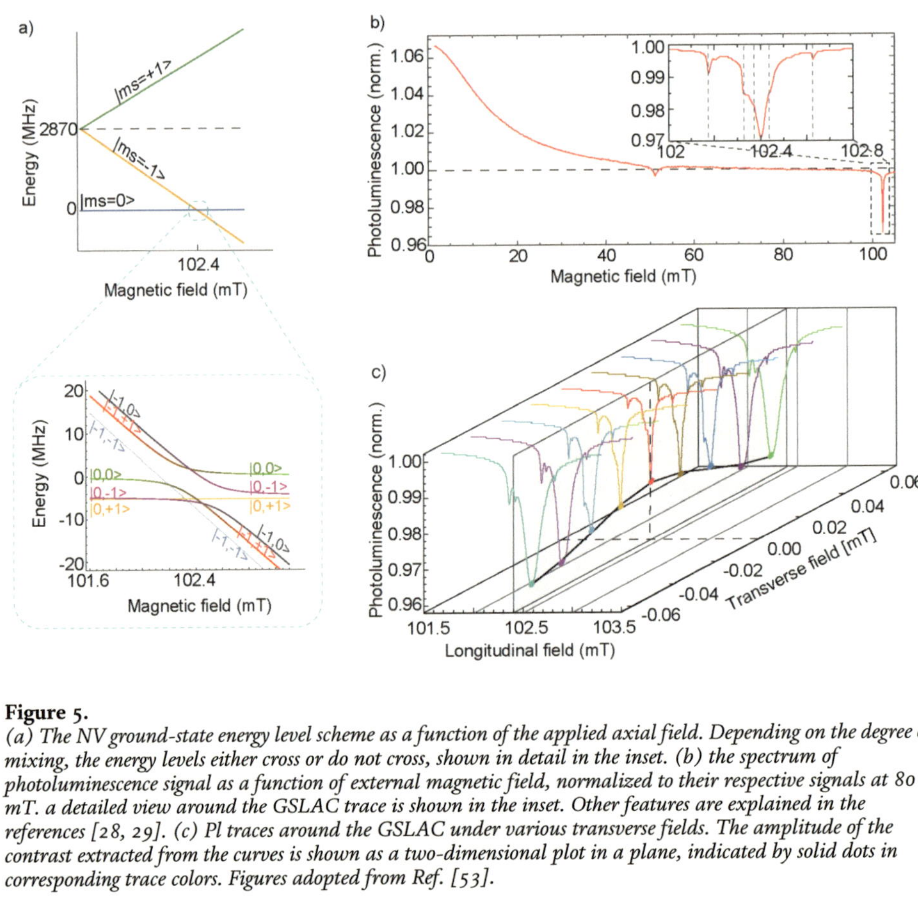

Figure 5.
(a) The NV ground-state energy level scheme as a function of the applied axial field. Depending on the degree of mixing, the energy levels either cross or do not cross, shown in detail in the inset. (b) the spectrum of photoluminescence signal as a function of external magnetic field, normalized to their respective signals at 80 mT. a detailed view around the GSLAC trace is shown in the inset. Other features are explained in the references [28, 29]. (c) Pl traces around the GSLAC under various transverse fields. The amplitude of the contrast extracted from the curves is shown as a two-dimensional plot in a plane, indicated by solid dots in corresponding trace colors. Figures adopted from Ref. [53].

applied transverse field is shown in **Figure 6(b)**. This is then demodulated by a LIA which delivers the information of both the amplitude and the angle of the magnetic field to be measured, as shown in **Figure 6(c)**. The reference phase of the LIA was set so that a magnetic field along the x-axis corresponds to phase zero (and negative amplitude). The LIA output shows a maximum value at 0° when applying a field along x-axis, shown in **Figure 6(c)**. An applied field in any other direction leads to an oscillating PL signal with a corresponding phase. Therefore, the phase output of the LIA is the angle between the transverse field to be measured and the defined x and y axes.

All Cartesian magnetic-field components can be directly read out in real time with equal sensitivity in all directions, with the above mentioned x and y plus the previously demonstrated longitudinal magnetic field measurement [25]. Note here, that the coordinate axes in the x-y plane is set by the reference phase of the LIA for the transverse-field signal demodulation while the phase for z-axis demodulation is tuned to maximize the amplitude of the response signal.

The experimental setup is based on **Figure 1** and extended with three pairs of orthogonal Helmholtz coils wound on a 3D printed mount. Sinusoidal signals for field modulations in the longitudinal and transverse directions and references for

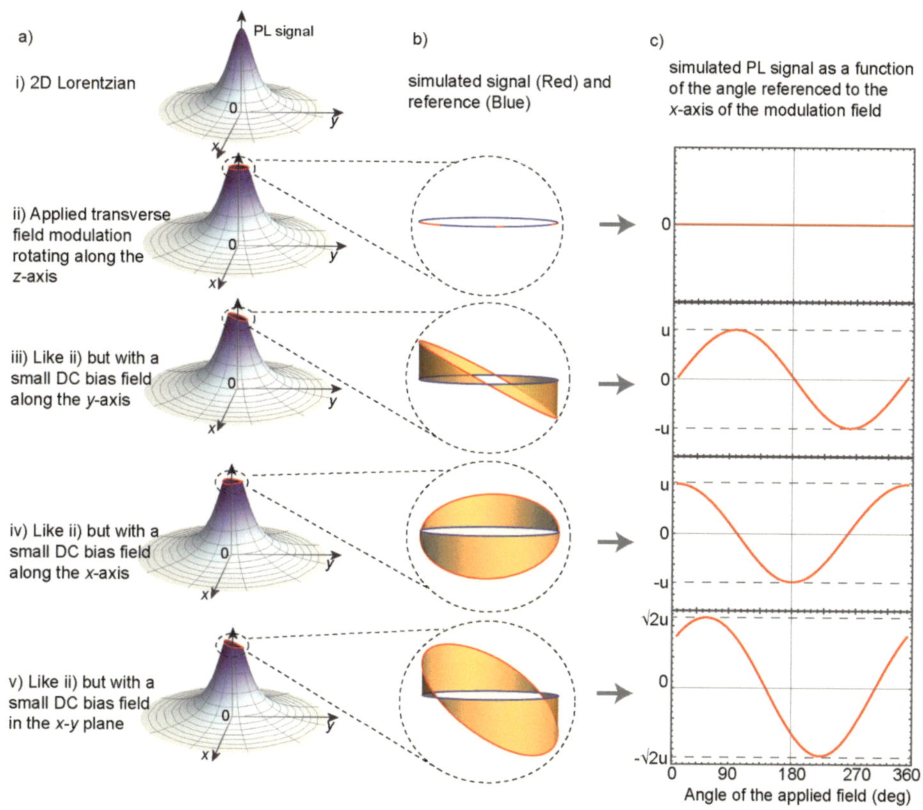

Figure 6.
Principle of the vectorial magnetic-field sensing. (a) Simulated photoluminescence signals as a function of transverse field. The trajectories of PL signals superimposed on the 2-D Lorentzian contrast function are indicated with red curves. Figures i) to v), correspond to the following cases: i) no modulated fields and no bias fields, ii) with modulated fields but no bias fields (the time-averaged PL drops), iii) with modulated fields and bias field along y, iv) with modulated fields and bias field along x, v) with modulated fields and bias field in x-y plane, respectively. (b) the subtraction of photoluminescence signals with/without presence (red curve/blue curve) of a bias field. (c) Simulated analog photoluminescence signals as a function of the angle of the modulated field referenced to the x-axis for transverse fields, flipped by 180° corresponding to (a) and (b). Figures adopted from Ref. [53].

the demodulation by two LIAs are provided by a two-channel function generator. One of the signals is split into two with one of them further passing through a phase shifter. These two signals are applied to two pairs of the Helmholtz coils with the same frequency but 90-degree-shifted relative phase (along the x and y axes). The Helmholtz coil pairs for the field modulation can also be used to calibrate the response to AC and DC magnetic fields $\mathbf{B} = (B_x, B_y, B_z)$. The applied fields, in the range of $\pm 4\ \mu T$ along each direction, are calibrated by flux gate magnetometers and

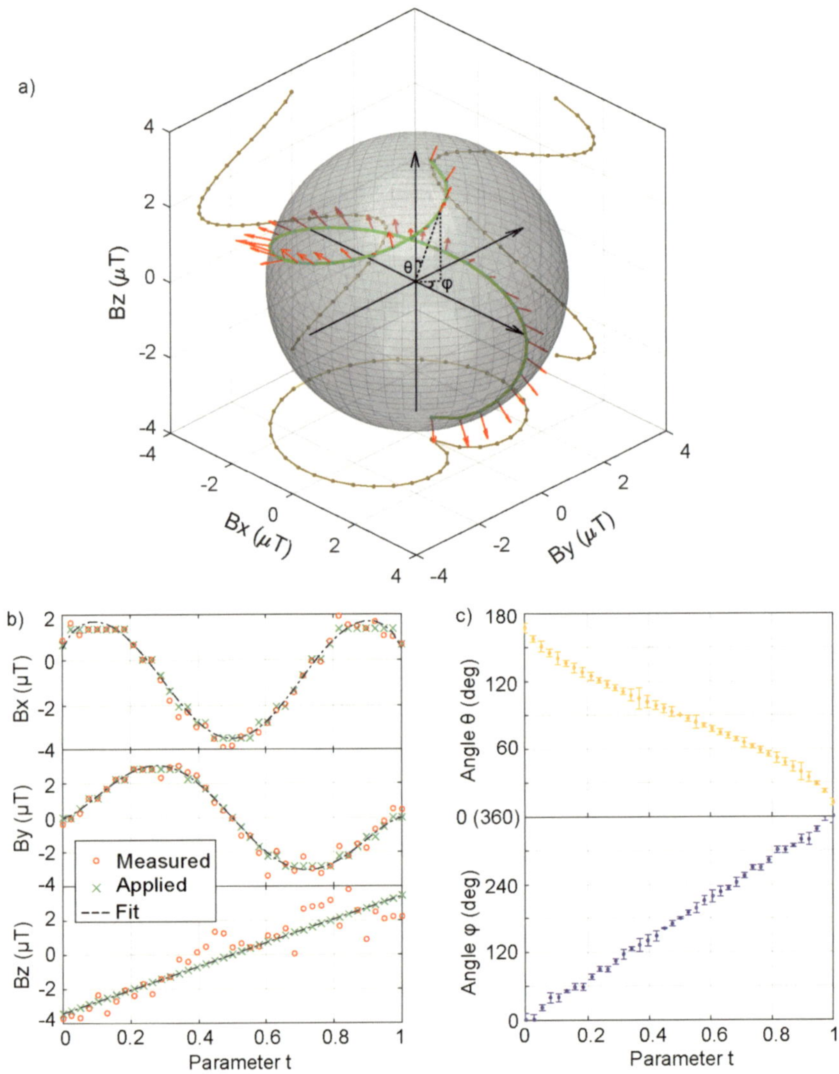

Figure 7.
Demonstration of full vector sensing capability. (a) Trajectory of the detected magnetic fields using the microwave-free vector magnetometer. The green curve indicates the 3D applied field and the brown curve is the projection on x-y, x-z, y-z planes. The vectors of the measured fields are indicated by red arrows. (b) the decomposed Cartesian components of both the measured (red circles) and applied (green points) magnetic fields. The applied field follows a parametric curve (black dashed lines) with $B_x = \sqrt{|B|^2 - Bz^2}\cos(2\pi t)\mu T$, $B_y = \sqrt{|B|^2 - Bz^2}\sin(2\pi t)\mu T$ *and* $B_z = 6.82t - 3.41\mu T$. *(c) the altitude angle (between* **B** *and the z-axis)* $\theta = \arccos(B_z/|B|)$ *(yellow dots) and the azimuth angle (between the projection of* **B** *in the x-y plane and the x-axis)* $\phi = \arctan(B_y/B_x) = 2\pi t$ *(blue dots) for each measured point. In the experiment, the altitude angle* θ *decreases in time from* 180° *to* 0° *and the azimuth angle* ϕ *increases from* 0° *to* 360°. *Figures adopted from Ref. [53].*

consistent with *a priori* calculations from the known coil geometry and the applied currents.

As a demonstration of full-vector sensing capacity, a set of static magnetic field vectors, whose trajectories were designed along a 3D a spiral curve on a sphere, was applied and measured. **Figure 7(a)** shows the spiral curve and the corresponding amplitudes are displayed in **Figure 7(b)**. The component in z direction is $|B| \cos\theta$, where $|B|$ is the magnitude of the applied magnetic field vectors and θ is the altitude angle (between the magnetic field to be measured and the z-axis). The B_x and B_y are $|B| \sin\theta \cos\phi$ and $|B| \sin\theta \sin\phi$, respectively, where ϕ is the azimuth angle (between the projection of **B** in the x-y plane and the x-axis). **Figure 7(b)** shows the corresponding values of B_x and B_y. The measured field components in x and y directions show good agreement with the amplitudes determined by *a priori* calculations. The trajectory was measured multiple times and the angles were reconstructed every time, shown in **Figure 7(c)** with the statistical error.

With the basic protocol established, this simultaneous vector magnetometry method should be extendable to single-NV probes. Single NV centers in diamond have been exploited to detect fluctuating magnetic fields (used as scalar relaxometry magnetometers) without microwaves [53, 54] as they share the spin dynamics near the GSLAC investigated above and in literature [53]. Both techniques, the present GSLAC-based vector magnetometry and relaxometry magnetometry, rely on monitoring the photoluminescence signal when the NVs are precisely turned to/near the GSLAC. Since they are operated with similar apparatus, the experimental setup for the latter can be extended for vector magnetic sensing by adding a set of 3D Helmholtz coil pairs and two LIAs. As there are no technical barriers for implementing the protocol onto a single NV center, it appears realistic to achieve nanoscale vector magnetometry applicable in a broad range of temperatures.

3. Zero-field magnetometry

Negatively charged NV centers in diamond have garnered wide interest as magnetometers [1, 10, 15, 25, 51, 55], with diverse applications ranging from electron spin resonance and biophysics to material science [56–62]. However, typical operation of an NV magnetometer requires an applied bias magnetic field to nonambiguously resolve magnetically sensitive features in the level structure. Due to the Zeeman effect, the bias field lifts degeneracy among the NV ground-state magnetic sublevels, allowing microwave transitions between spin states to be addressed individually [11]. Such a bias field can be detrimental for applications where it could perturb the system under examination, such as measurements in magnetically shielded environments and in magnetic susceptometry [63].

Elimination of the need for a bias field would extend the dynamic range of NV-based magnetic sensors to zero field. Zero-field NV magnetometry opens up new application avenues, and makes these versatile, solid-state sensors competitive with other magnetic field probes such as SQUIDs and alkali-vapor magnetometers [64, 65], because, despite the lower sensitivity of NVs, they offer additional benefits due to high spatial resolution and small sensor size, capability of being operated over large ranges of pressure and temperature, and wide bandwidth [1]. The relative simplicity of NVs magnetometers can readily complement existing sensors in applications such as tracking field fluctuations in experimental searches for electric dipole moments [66], zero- and ultra low-field NMR [67, 68], and magnetoencelography or magnetocardiography [26, 69].

Magnetically sensitive microwave transitions within NV centers can be probed using the ODMR technique. At zero field, however, these transitions overlap, and shift equally with opposite sign in response to magnetic fields. Therefore, NV ensembles have been considered unusable as zero-field magnetometers [1], except in certain cases, for detecting ac fields in the presence of applied microwaves [70].

A schematic of the experimental apparatus is described in Zheng *et al.* [71]. The ODMR traces around zero field are taken by sweeping the frequency of linear microwaves under a range of magnetic field values around zero. The ODMR acquired spectra are presented in **Figure 8(a)**, which show transitions originating from all crystal orientations, with hyperfine structure resolved for nuclear spin I = 1. These transitions are shown in detail at specific field values in **Figure 8(b)**, including at zero field, where 12 transitions overlap and merge into four distinct features.

The authors employ circularly-polarized microwaves to drive a single electron spin transition out of the feature composed of overlapping resonances. This single transition, due to Zeeman effect, has a linear dependence on the external magnetic field. The circular microwaves are generated by two, 200-μm wires separated by a distance $d = 4.5$ mm, printed on a circuit board that follows the design of Ref. [72]. The diamond is placed above the board with $d/2$ distance. Each wire carries a MW signal split from the same source, with one passed through a variable phase shifter. This design and arrangement yield orthogonal oscillating magnetic fields to the $\langle 111 \rangle$-oriented NV centers in the diamond sample, which is verified with the high contrast between ODMR traces at certain field with σ^+ or σ^- circularly-polarized microwaves (**Figure 9(a)**). The microwave field can be varied between linear and circular (**Figure 9(b)**).

Figure 9 shows the efficacy of the circular MW polarizing. In particular, **Figure 9(b)** and **(c)** demonstrate a relative suppression of σ^+ and σ^- transitions to below 1% of the maximum contrast, respectively. Previously overlapping transitions are thus isolated, removing the symmetric dependence as a function of the field to be measured.

It is demonstrated with a 250 pT/$\sqrt{\text{Hz}}$ noise floor for the above NV-based zero-field magnetometer. The device employs NV ensemble in diamond with a well-

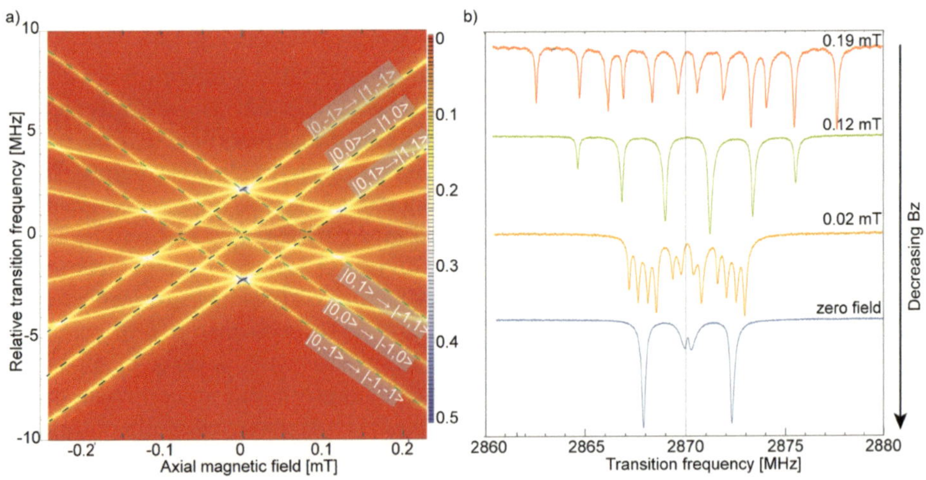

Figure 8.
(a) ODMR spectra with linearly polarized MWs as a function of the axial magnetic field, with transitions originating from all crystal-axis orientations. Those transitions corresponding to NVs oriented along the direction of the applied field are labeled ($|m_s, m_I\rangle$) and overlaid with the calculated transitions. The transitions that are not labeled are assigned to NV centers that are not oriented along the $\langle 111 \rangle$ direction. These features overlap because they experience a common relative angle with the applied field. (b) ODMR traces for selected values of B_z. The central transitions are split at near zero field. The higher-energy peak at ≈ 2872 MHz are the transitions $|m_s = 0, m_I = -1\rangle \rightarrow |m_s = +1, m_I = -1\rangle$ and $|m_s = 0, m_I = +1\rangle \rightarrow |m_s = -1, m_I = +1\rangle$. The lower-energy peak at ≈ 2868 MHz corresponds to the transitions $|m_s = 0, m_I = +1\rangle \rightarrow |m_s = +1, m_I = +1\rangle$ and $|m_s = 0, m_I = -1\rangle \rightarrow |m_s = -1, m_I = -1\rangle$. Figures adopted from Ref. [71].

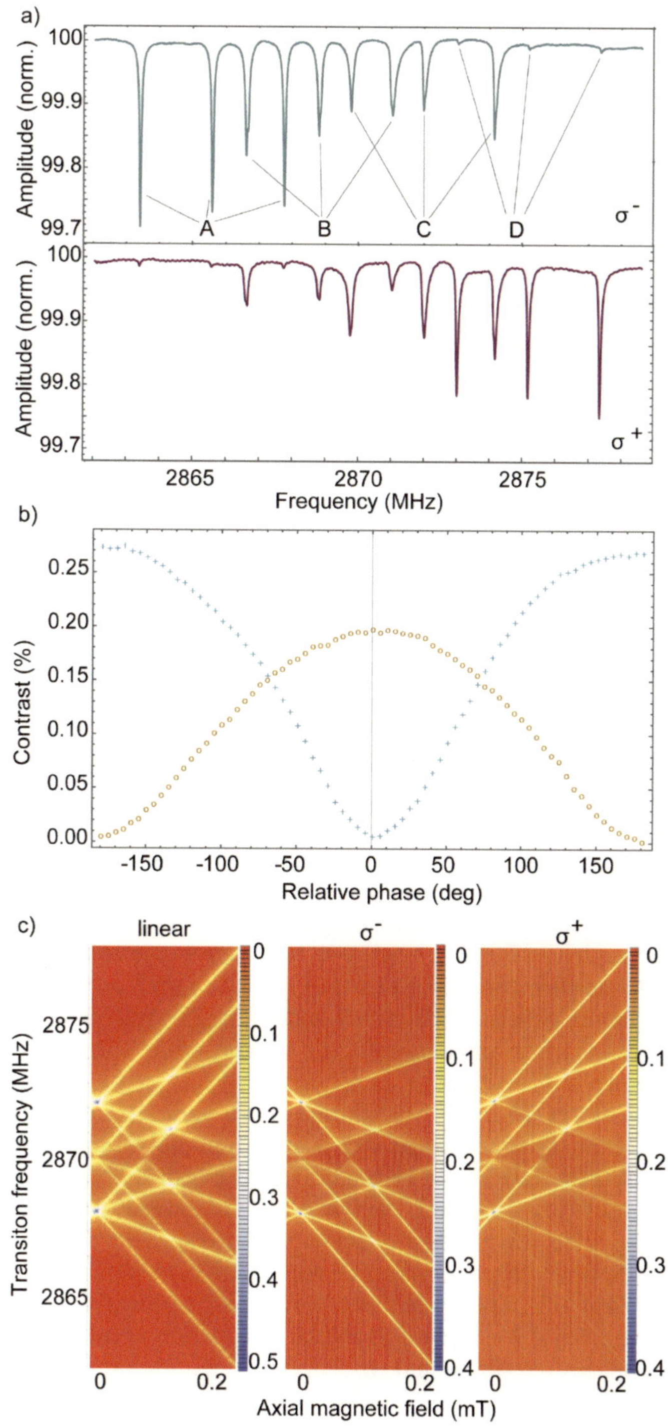

Figure 9.
(a) ODMR spectrum at certain field with circularly-polarized MWs (the polarization of the applied MW is indicated at the right bottom corner of each sub-figure: Top σ^- and bottom σ^+). The peaks are labeled into 4 groups, peaks A and D (B and C) correspond to the transitions from $|m_s = 0\rangle$ to $|m_s = -1\rangle$ and $|m_s = 0\rangle$ to $|m_s = 1\rangle$ of on-axis (off-axis) NVs, respectively. (b) Subtracted contrasts of A and D by Lorentzian fitting, as a function of the relative phase between two applied microwave fields. Blue crosses (amber circles) indicate the amplitudes of A (D). (c) ODMR traces under linear, σ^-, σ^+ MWs at various magnetic fields. The color scale indicates the peak depth, in percent, relative to the off-resonant case. Figures adopted from Ref. [71].

resolved ground-state hyperfine structure, and uses circular microwaves to selectively excite transitions between the magnetic sublevels, in the absence of such selectivity, yield ODMR signals that are first-order magnetically insensitive at near zero field. This device can find use in applications where a bias field is undesirable and extends the dynamic range of NV magnetometry to cover existing zero-field technologies such as SQUIDs and alkali-vapor magnetometers. Further improvements in the present technique will result in sensitivities that are beneficial for NMR at zero and ultra-low fields and, with further miniaturization, these zero-field diamond sensors can be useful in biomagnetic applications such as magnetocardiography and magnetoencelography.

4. Outlook

This chapter focuses on a number of novel magnetic sensing techniques with NV centers in diamond. In Section 2, a novel method for a microwave-free magnetometer based on the GSLAC, together with its utilization to achieve eddy-current imaging, is described. In addition, approaches to improve the sensitivity of the MW-free magnetometers and a different detection method–cavity-enhanced singlet-absorption measurement are discussed. In Section 3, we introduce a vector magnetometer based on the GSLAC microwave-free magnetic sensing technique, which enables simultaneous measurement of all Cartesian components of a magnetic field. In Section 4, we presented a magnetometric prototype based on ODMR which is realized with circularly polarized MWs. This magnetometer operates at zero ambient field and thus extends magnetic sensing dynamic range and opens up new application avenues. All the above magnetometers are realized in continuous-wave mode and can be potentially operated with pulse sequences.

This chapter advances in ultra-sensitive, high-bandwidth magnetometry with NV ensembles. Significant work remains to bring each to its full potential. Advances will most likely be driven by both improved pulse sequences and better materials. Additionally, the performance can be further improved with approaches such as increasing the photon collection efficiency, extending NV spin coherence times, increasing readout fidelity, and suppressing common-mode noise by differential detection and so on.

The ability to detect magnetic field patterns with high magnetic sensitivity and spatial resolution could prove a useful characterization tool in a number of fields. In particular, NV magnetometers can be used for non-invasively magnetically imaging biomagnetic systems (e.g., neurons, cardiac cells, and magnetic organs used for navigation) using microwave-free probes, vectorial stray-field imaging of magnetic structures, and detecting nuclear magnetic resonance of chemically exchanging systems at zero field [73]. The presented techniques are potentially applicable to single NV center probes, which would facilitate extraction of magnetic information with nanoscale spatial resolution and would boost numerous applications.

Additional information

This chapter is based on the PhD Thesis of Huijie Zheng.

Author details

Huijie Zheng[1*], Arne Wickenbrock[1,2], Georgios Chatzidrosos[1], Lykourgos Bougas[1,2], Nathan Leefer[1], Samer Afach[1], Andrey Jarmola[3], Victor M. Acosta[4], Jingyan Xu[5], Geoffrey Z. Iwata[2], Till Lenz[1], Zhiyin Sun[6], Chen Zhang[7], Takeshi Ohshima[8], Hitoshi Sumiya[9], Kazuo Nakamura[10], Junichi Isoya[11], Jörg Wrachtrup[7] and Dmitry Budker[1,2,3]

1 Johannes Gutenberg-Universität Mainz, Mainz, Germany

2 Helmholtz Institut Mainz, Mainz, Germany

3 Department of Physics, University of California, Berkeley, CA, USA

4 Department of Physics and Astronomy, University of New Mexico, Center for High Technology Materials, Albuquerque, NM, USA

5 Chinese Academy of Sciences, Key Lab of Quantum Information, University of Science and Technology of China, Hefei, People's Republic of China

6 Laboratory for Space Environment and Physical Sciences, Harbin Institute of Technology, Harbin, China

7 Institute of Physics, University of Stuttgart and Institute for Quantum Science and Technology IQST, Stuttgart, Germany

8 Takasaki Advanced Radiation Research Institute, National Institutes for Quantum and Radiological Science and Technology, Takasaki, Japan

9 Advanced Materials Laboratory, Sumitomo Electric Industries, Ltd., Itami, Japan

10 Application Technology Research Institute, Tokyo Gas Company, Ltd., Yokohama, Japan

11 Faculty of Pure and Applied Sciences, University of Tsukuba, Tsukuba, Japan

*Address all correspondence to: zheng@uni-mainz.de

References

[1] L Rondin, J-P Tetienne, T Hingant, J-F Roch, P Maletinsky, and V Jacques. Magnetometry with nitrogen-vacancy defects in diamond. *Reports on Progress in Physics*, 77(5):056503, 2014.

[2] G. Kucsko, P. C. Maurer, N. Y. Yao, M. Kubo, H. J. Noh, P. K. Lo, H. Park, and M. D. Lukin. Nanometre-scale thermometry in a living cell. *Nature Physics Letter*, 500:54–58, 2013.

[3] V. M. Acosta, E. Bauch, M. P. Ledbetter, A. Waxman, L.-S. Bouchard, and D. Budker. Temperature dependence of the nitrogen-vacancy magnetic resonance in diamond. *Phys. Rev. Lett.*, 104:070801, 2010.

[4] F. Dolde, H. Fedder, M. W. Doherty, T. Nobauer, F. Rempp, G. Balasubramanian, T. Wolf, F. Reinhard, L. C. L. Hollenberg, F. Jelezko, and J. Wrachtrup. Electric-field sensing using single diamond spins. *Nature Physics Letter*, 455:644–647, 2008.

[5] Preeti Ovartchaiyapong, Kenneth W. Lee, Bryan A. Myers, Ania C. Bleszynski Jayich, A. Stacey, D. Budker, and L. C. L. Hollenberg. Dynamic strain-mediated coupling of a single diamond spin to a mechanical resonator. *Nat Commun*, 5, 2014.

[6] M. P. Ledbetter, K. Jensen, R. Fischer, A. Jarmola, and D. Budker. Gyroscopes based on nitrogen-vacancy centers in diamond. *Phys. Rev. A*, 86: 052116, 2012.

[7] Ashok Ajoy and Paola Cappellaro. Stable three-axis nuclear-spin gyroscope in diamond. *Phys. Rev. A*, 86:062104, 2012.

[8] D. Maclaurin, M. W. Doherty, L. C. L. Hollenberg, and A. M. Martin. Measurable quantum geometric phase from a rotating single spin. *Phys. Rev. Lett.*, 108:240403, 2012.

[9] Eric van Oort, Paul Stroomer, and Max Glasbeek. Low-field optically detected magnetic resonance of a coupled triplet-doublet defect pair in diamond. *Phys. Rev. B*, 42:8605–8608, 1990.

[10] Thomas Wolf, Philipp Neumann, Kazuo Nakamura, Hitoshi Sumiya, Takeshi Ohshima, Junichi Isoya, and Jörg Wrachtrup. Subpicotesla diamond magnetometry. *Phys. Rev. X*, 5:041001, 2015.

[11] J. M. Taylor, P. Cappellaro, L. Childress, L. Jiang, Philipp Neumann, D. Budker, P. R. Hemmer, A. Yacoby, R. Walsworth, and M. D. Lukin. High-sensitivity diamond magnetometer with nanoscale resolution. *Nature Physics*, 4: 810–816, 2008.

[12] VM Acosta, E Bauch, A Jarmola, LJ Zipp, MP Ledbetter, and D Budker. Broadband magnetometry by infrared-absorption detection of nitrogen-vacancy ensembles in diamond. *Applied Physics Letters*, 97(17):174104, 2010.

[13] V. M. Acosta, A. Jarmola, E. Bauch, and D. Budker. Optical properties of the nitrogen-vacancy singlet levels in diamond. *Phys. Rev. B*, 82:201202, 2010.

[14] Y. Dumeige, M. Chipaux, V. Jacques, F. Treussart, J.-F. Roch, T. Debuisschert, V. M. Acosta, A. Jarmola, K. Jensen, P. Kehayias, and D. Budker. Magnetometry with nitrogen-vacancy ensembles in diamond based on infrared absorption in a doubly resonant optical cavity. *Phys. Rev. B*, 87:155202, 2013.

[15] K. Jensen, N. Leefer, A. Jarmola, Y. Dumeige, V. M. Acosta, P. Kehayias, B. Patton, and D. Budker. Cavity-enhanced room-temperature magnetometry using absorption by nitrogen-vacancy centers in diamond. *Phys. Rev. Lett.*, 112:160802, 2014.

[16] Arne Wickenbrock, Sarunas Jurgilas, Albert Dow, Luca Marmugi, and Ferruccio Renzoni. Magnetic induction tomography using an all-optical 87rb atomic magnetometer. *Opt. Lett.*, 39(22):6367–6370, 2014.

[17] Cameron Deans, Luca Marmugi, Sarah Hussain, and Ferruccio Renzoni. Electromagnetic induction imaging with a radio-frequency atomic magnetometer. *Applied Physics Letters*, 108(10):103503, 2016.

[18] Arne Wickenbrock, Nathan Leefer, John W Blanchard, and Dmitry Budker. Eddy current imaging with an atomic radio-frequency magnetometer. *Applied Physics Letters*, 108(18):183507, 2016.

[19] Georgios Chatzidrosos, Arne Wickenbrock, Lykourgos Bougas, Huijie Zheng, Oleg Tretiak, Yu Yang, and Dmitry Budker. Eddy-current imaging with nitrogen-vacancy centers in diamond. *Phys. Rev. Applied*, 11:014060, Jan 2019.

[20] David A Simpson, Jean-Philippe Tetienne, Julia M McCoey, Kumaravelu Ganesan, Liam T Hall, Steven Petrou, Robert E Scholten, and Lloyd CL Hollenberg. Magneto-optical imaging of thin magnetic films using spins in diamond. *Scientific reports*, 6(1):1–8, 2016.

[21] J-P Tetienne, L Rondin, P Spinicelli, M Chipaux, T Debuisschert, J-F Roch, and V Jacques. Magnetic-field-dependent photodynamics of single nv defects in diamond: an application to qualitative all-optical magnetic imaging. *New Journal of Physics*, 14(10):103033, 2012.

[22] P. Maletinsky, S. Hong, M. S. Grinolds, B. Hausmann, M. D. Lukin, R. L. Walsworth, M. Loncar, and A. Yacoby. A robust scanning diamond sensor for nanoscale imaging with single nitrogen-vacancy centres. *Nat Nano*, 7 (5):320–324, 2012.

[23] L. Rondin, J.-P. Tetienne, P. Spinicelli, C. Dal Savio, K. Karrai, G. Dantelle, A. Thiaville, S. Rohart, J.-F. Roch, and V. Jacques. Nanoscale magnetic field mapping with a single spin scanning probe magnetometer. *Applied Physics Letters*, 100(15), 2012.

[24] Jared H Cole and Lloyd C L Hollenberg. Scanning quantum decoherence microscopy. *Nanotechnology*, 20(49):495401, 2009.

[25] Arne Wickenbrock, Huijie Zheng, Lykourgos Bougas, Nathan Leefer, Samer Afach, Andrey Jarmola, Victor M Acosta, and Dmitry Budker. Microwave- free magnetometry with nitrogen-vacancy centers in diamond. *Applied Physics Letters*, 109, 2016.

[26] Riccardo Fenici, Donatella Brisinda, and Anna Maria Meloni. Clinical application of magnetocardiography. *Expert Review of Molecular Diagnostics*, 5 (3):291–313, 2005.

[27] Seiji Armstrong, Lachlan J. Rogers, Roger L. McMurtrie, and Neil B. Manson. Nv-nv electron-electron spin and nv-ns electron-electron and electron-nuclear spin interaction in diamond. *Physics Procedia*, 3(4):1569–1575, 2010.

[28] LT Hall, P Kehayias, DA Simpson, A Jarmola, A Stacey, D Budker, and LCL Hollenberg. Detection of nanoscale electron spin resonance spectra demonstrated using nitrogen-vacancy centre probes in diamond. *Nature communications*, 7(1):1–9, 2016.

[29] James D. A. Wood, David A. Broadway, Liam T. Hall, Alastair Stacey, David A. Simpson, Jean-Philippe Tetienne, and Lloyd C. L. Hollenberg. Wide-band nanoscale magnetic resonance spectroscopy using quantum relaxation of a single spin in diamond. *Phys. Rev. B*, 94:155402, 2016.

[30] Ivády, Viktor and Zheng, Huijie and Wickenbrock, Arne and Bougas,

Lykourgos and Chatzidrosos, Georgios and Nakamura, Kazuo and Sumiya, Hitoshi and Ohshima, Takeshi and Isoya, Junichi and Budker, Dmitry and others. Photoluminescence at the ground state level anticrossing of the nitrogen-vacancy center in diamond. 2020:arXiv preprint arXiv:2006.05085

[31] Corey J Cochrane, Jordana Blacksberg, Mark A Anders, and Patrick M Lenahan. Vectorized magnetometer for space applications using electrical readout of atomic scale defects in silicon carbide. *Scientific Reports,* 6:37077, 2016.

[32] James Lenz and S Edelstein. Magnetic sensors and their applications. *IEEE Sensors Journal,* 6(3):631–649, 2006.

[33] D Le Sage, K Arai, DR Glenn, SJ DeVience, LM Pham, L Rahn-Lee, MD Lukin, A Yacoby, A Komeili, and RL Walsworth. Optical magnetic imaging of living cells. *Nature,* 496(7446):486–489, 2013.

[34] Matti Hämäläinen, Riitta Hari, Risto J. Ilmoniemi, Jukka Knuutila, and Olli V. Lounasmaa. Magnetoencephalography—theory, instrumentation, and applications to noninvasive studies of the working human brain. *Rev. Mod. Phys.,* 65:413–497, Apr 1993.

[35] Luong Van Su, Jeng Jen-Tzong, Lu Chih-Cheng, and Hsu Hua-Yi. Low-noise tunneling-magnetoresistance vector magnetometers with flux chopping technique. *Measurement,* 109: 297–303, 2017.

[36] T. Schönau, V. Zakosarenko, M. Schmelz, R. Stolz, S. Anders, S. Linzen, M. Meyer, and H.-G. Meyer. A three-axis squid-based absolute vector magnetometer. *Review of Scientific Instruments,* 86(10):105002, 2015.

[37] S. J. Seltzer and M. V. Romalis. Unshielded three-axis vector operation of a spin-exchange-relaxation-free atomic magnetometer. *Applied Physics Letters,* 85(20):4804–4806, 2004.

[38] B. Patton, E. Zhivun, D. C. Hovde, and D. Budker. All-optical vector atomic magnetometer. *Phys. Rev. Lett.,* 113: 013001, July 2014.

[39] Georgios Chatzidrosos, Arne Wickenbrock, Lykourgos Bougas, Nathan Leefer, Teng Wu, Kasper Jensen, Yannick Dumeige, and Dmitry Budker. Miniature cavity-enhanced diamond magnetometer. *Phys. Rev. Applied*, 8:044019, Oct 2017.

[40] Y. Dumeige, M. Chipaux, V. Jacques, F. Treussart, J.-F. Roch, T. Debuisschert, V. M. Acosta, A. Jarmola, K. Jensen, P. Kehayias, and D. Budker. Magnetometry with nitrogen-vacancy ensembles in diamond based on infrared absorption in a doubly resonant optical cavity. *Phys. Rev. B*, 87:155202, Apr 2013.

[41] Cheng Gong, Lin Li, Zhenglu Li, Huiwen Ji, Alex Stern, Yang Xia, Ting Cao, Wei Bao, Chenzhe Wang, Yuan Wang, et al. Discovery of intrinsic ferromagnetism in two-dimensional van der waals crystals. *Nature*, 546(7657): 265, 2017.

[42] Manuel Bonilla, Sadhu Kolekar, Yujing Ma, Horacio Coy Diaz, Vijaysankar Kalappattil, Raja Das, Tatiana Eggers, Humberto R Gutierrez, Manh-Huong Phan, and Matthias Batzill. Strong room-temperature ferromagnetism in vse 2 monolayers on van der waals substrates. *Nature nanotechnology*, 13(4):289, 2018.

[43] Matthew Pelliccione, Alec Jenkins, Preeti Ovartchaiyapong, Christopher Reetz, Eve Emmanouilidou, Ni Ni, and Ania C Bleszynski Jayich. Scanned probe imaging of nanoscale magnetism at cryogenic temperatures with a single-spin quantum sensor. *Nature nanotechnology,* 11(8):700, 2016.

[44] Isabell Gross, W Akhtar, V Garcia, LJ Martínez, Saddem Chouaieb, K Garcia, C Carrétéro, A Barthélémy, P Appel, P Maletinsky, et al. Real-space imaging of non-collinear antiferromagnetic order with a single-spin magnetometer. *Nature*, 549(7671): 252, 2017.

[45] Jennifer M. Schloss, John F. Barry, Matthew J. Turner, and Ronald L. Walsworth. Simultaneous broadband vector magnetometry using solid-state spins. *Phys. Rev. Applied,* 10:034044, Sep 2018.

[46] Chen Zhang, Heng Yuan, Ning Zhang, Lixia Xu, Jixing Zhang, Bo Li, and Jiancheng Fang. Vector magnetometer based on synchronous manipulation of nitrogen-vacancy centers in all crystal directions. *Journal of Physics D: Applied Physics*, 51(15): 155102, 2018.

[47] Yi-Xiang Liu, Ashok Ajoy, and Paola Cappellaro. Nanoscale vector dc magnetometry via ancilla-assisted frequency up-conversion. *Phys. Rev. Lett.,* 122:100501, Mar 2019.

[48] Pengfei Wang, Zhenheng Yuan, Pu Huang, Xing Rong, Mengqi Wang, Xiangkun Xu, Changkui Duan, Chenyong Ju, Fazhan Shi, and Jiangfeng Du. High-resolution vector microwave magnetometry based on solid-state spins in diamond. *Nature communications,* 6(1):1–5, 2015.

[49] Lucas Thiel, Zhe Wang, Märta A Tschudin, Dominik Rohner, Ignacio Gutiérrez-Lezama, Nicolas Ubrig, Marco Gibertini, Enrico Giannini, Alberto F Morpurgo, and Patrick Maletinsky. Probing magnetism in 2d materials at the nanoscale with single-spin microscopy. *Science,* 364(6444): 973–976, 2019.

[50] Huijie Zheng, Zhiyin Sun, Georgios Chatzidrosos, Chen Zhang, Kazuo Nakamura, Hitoshi Sumiya, Takeshi Ohshima, Junichi Isoya, Jörg Wrachtrup, Arne Wickenbrock, and Dmitry Budker. Microwave-free vector magnetometry with nitrogen-vacancy centers along a single axis in diamond. *Phys. Rev. Applied,* 13:044023, Apr 2020.

[51] Huijie Zheng, Georgios Chatzidrosos, Arne Wickenbrock, Lykourgos Bougas, Reinis Lazda, Andris Berzins, Florian Helmuth Gahbauer, Marcis Auzinsh, Ruvin Ferbe, and Dmitry Budker. Level anti-crossing magnetometry with color centers in diamond. *Proc. of SPIE,* page 101190X, Feb 2017.

[52] R. J. Epstein, F. M. Mendoza, Y. K. Kato, and D. D. Awschalom. Anisotropic interactions of a single spin and dark-spin spectroscopy in diamond. *Nature Physics,* 1:94–98, 2005.

[53] David A. Broadway, James D. A. Wood, Liam T. Hall, Alastair Stacey, Matthew Markham, David A. Simpson, Jean-Philippe Tetienne, and Lloyd C. L. Hollenberg. Anticrossing spin dynamics of diamond nitrogen-vacancy centers and all-optical low-frequency magnetometry. *Phys. Rev. Applied*, 6: 064001, Dec 2016.

[54] James DA Wood, Jean-Philippe Tetienne, David A Broadway, Liam T Hall, David A Simpson, Alastair Stacey, and Lloyd CL Hollenberg. Microwave-free nuclear magnetic resonance at molecular scales. *Nature communications,* 8(1):1–6, 2017.

[55] Aharon Blank, Guy Shapiro, Ran Fischer, Paz London, and David Gershoni. Optically detected magnetic resonance imaging. *Applied Physics Letters,* 106(3):034102, 2015.

[56] Gopalakrishnan Balasubramanian, IY Chan, Roman Kolesov, Mohannad Al-Hmoud, Julia Tisler, Chang Shin, Changdong Kim, Aleksander Wojcik, Philip R Hemmer, Anke Krueger, et al. Nanoscale imaging magnetometry with

diamond spins under ambient conditions. *Nature*, 455(7213):648, 2008.

[57] F. Jelezko and J. Wrachtrup. Single defect centres in diamond: A review. *physical status solidi (a)*, 203(13):3207–3225, 2006.

[58] F Kong, P.J Zhao, X.Y. Ye, Z.C Wang, Z.Y Qin, P. Yu, J.H. Su, F.Z. Shi, and J.F. Du. Nanoscale zero-field electron spin resonance spectroscopy. *Nature Communications*, 9:1563, Apr 2018.

[59] John F. Barry, Matthew J. Turner, Jennifer M. Schloss, David R. Glenn, Yuyu Song, Mikhail D. Lukin, Hongkun Park, and Ronald L. Walsworth. Optical magnetic detection of single-neuron action potentials using quantum defects in diamond. *Proceedings of the National Academy of Sciences*, 113(49):14133–14138, 2016.

[60] Steffen Steinert, Florian Dolde, Philipp Neumann, Andrew Aird, Boris Naydenov, Gopalakrishnan Balasubramanian, Fedor Jelezko, and Joerg Wrachtrup. High sensitivity magnetic imaging using an array of spins in diamond. *Review of scientific instruments*, 81(4):043705, 2010.

[61] C. L. Degen. Scanning magnetic field microscope with a diamond singlespin sensor. *Applied Physics Letters*, 92(24):243111, 2008.

[62] P. London, J. Scheuer, J.-M. Cai, I. Schwarz, A. Retzker, M. B. Plenio, M. Katagiri, T. Teraji, S. Koizumi, J. Isoya, R. Fischer, L. P. McGuinness, B. Naydenov, and F. Jelezko. Detecting and polarizing nuclear spins with double resonance on a single electron spin. *Phys. Rev. Lett.*, 111:067601, Aug 2013.

[63] D. Eberbeck, A. P. Astalan, K. Petersson, F. Wiekhorst, C. Bergemann, C. Johansson, U. Steinhoff, H. Richter, A. Krozer, and L. Trahms. Ac susceptometry and magnetorelaxometry for magnetic nanoparticle based biomolecule detection. In Jos Vander Sloten, Pascal Verdonck, Marc Nyssen, and Jens Haueisen, editors, *4th European Conference of the International Federation for Medical and Biological Engineering*, pages 2317–2321, Berlin, Heidelberg, 2009. Springer Berlin Heidelberg.

[64] Guzhi Bao, Arne Wickenbrock, Simon Rochester, Weiping Zhang, and Dmitry Budker. Suppression of the nonlinear zeeman effect and heading error in earth-field-range alkali-vapor magnetometers. *Phys. Rev. Lett.*, 120: 033202, Jan 2018.

[65] Wei Quan, Kai Wei, and Hairong Li. Precision measurement of magnetic field based on the transient process in a k-rb-21ne co-magnetometer. *Opt. Express*, 25(8):8470–8483, Apr 2017.

[66] G. Bison, V. Bondar, P. Schmidt-Wellenburg, A. Schnabel, and J. Voigt. Sensitive and stable vector magnetometer for operation in zero and finite fields. *Opt. Express*, 26(13):17350–17359, Jun 2018.

[67] Min Jiang, Teng Wu, John W. Blanchard, Guanru Feng, Xinhua Peng, and Dmitry Budker. Experimental benchmarking of quantum control in zero-field nuclear magnetic resonance. *Science Advances*, 4(6), 2018.

[68] M. P. Ledbetter, I. M. Savukov, D. Budker, V. Shah, S. Knappe, J. Kitching, D.J. Michalak, S. Xu, and A. Pines. Zero-field remote detection of nmr with a microfabricated atomic magnetometer. *Proceedings of the National Academy of Sciences of the United States of America*, 105(7):2286–2290, 2008.

[69] A I Ahonen, M S Hamalainen, M J Kajola, J E T Knuutila, P P Laine, O V Lounasmaa, L T Parkkonen, J T Simola, and C D Tesche. 122-channel squid instrument for investigating the magnetic signals from the human brain. *Physica Scripta*, 1993(T49A):198, 1993.

[70] Soya Saijo, Yuichiro Matsuzaki, Shiro Saito, Tatsuma Yamaguchi, Ikuya Hanano, Hideyuki Watanabe, Norikazu Mizuochi, and Junko Ishi-Hayase. Ac magnetic field sensing using continuous-wave optically detected magnetic resonance of nitrogen-vacancy centers in diamond. *Applied Physics Letters,* 113(8):082405, 2018.

[71] Huijie Zheng, Jingyan Xu, Geoffrey Z. Iwata, Till Lenz, Julia Michl, Boris Yavkin, Kazuo Nakamura, Hitoshi Sumiya, Takeshi Ohshima, Junichi Isoya, Jörg Wrachtrup, Arne Wickenbrock, and Dmitry Budker. Zero-field magnetometry based on nitrogen-vacancy ensembles in diamond. *Phys. Rev. Applied,* 11:064068, Jun 2019.

[72] M. Mrozek, J. Mlynarczyk, D. S. Rudnicki, and W. Gawlik. Circularly polarized microwaves for magnetic resonance study in the ghz range: Application to nitrogen-vacancy in diamonds. *Applied Physics Letters,* 107 (1):013505, 2015.

[73] Danila A Barskiy, Michael CD Tayler, Irene Marco-Rius, John Kurhanewicz, Daniel B Vigneron, Sevil Cikrikci, Ayca Aydogdu, Moritz Reh, Andrey N Pravdivtsev, Jan-Bernd Hövener, et al. Zero-field nuclear magnetic resonance of chemically exchanging systems. *Nature communications,* 10(1):1–9, 2019.

Section 3

Electrical

Chapter 3

Unique Surface Modifications on Diamond Thin Films

Vadali Venkata Satya Siva Srikanth

Abstract

Diamond thin films are touted to be excellent in surface-sensitive sensing, electro-mechanical systems, and electrochemical applications. However, these applications often entail patterned active surfaces and subtle chemical surface modifications. But due to diamond's intrinsic hardness and chemical inertness, surface patterning (using micro-machining and ion etching) and chemical surface modifications, respectively, are very difficult. In the case of surface patterning, it is even more challenging to obtain patterns during synthesis. In this chapter, the direct patterning of sub-wavelength features on diamond thin film surface using a femto-second laser, rapid thermal annealing as a means to prepare the diamond thin film surface as an efficient direct charge transfer SERS substrate (in metal/insulator/semiconductor (MIS) configuration), and implantation of $^{14}N^{+}$ ions into the surface and sub-surface regions for enhancing the electrical conductivity of diamond thin film to a certain depth (in MIS configuration) will be discussed encompassing the processing strategies and different post-processing characteristics.

Keywords: diamond, thin films, surface patterning, rapid thermal annealing, ion implantation

1. Introduction

Scientists working on diamonds realized long back that the unsurpassed properties of a diamond could be tapped for many applications only when synthesized in thin-film form. Since then, chemical vapor deposition (CVD) [1, 2] has been the most generally used technique to synthesize various diamond thin films for multiple applications. CVD of diamond involves the low-pressure growth of a thin film of a network of tetrahedrally bonded carbon atoms (on a diamond-seeds pre-treated substrate) from a suitable gas phase. CVD of diamond thin films typically involves activation of the appropriate gas mixtures (for example, a mixture of H_2 and CH_4 under low pressure activated by microwave energy), gas-phase reactions (leading to the formation of diamond forming gas species), diffusion of the diamond forming gas species onto the substrate and formation of a non-volatile thin layer of tetrahedrally bonded carbon atoms network. By controlling the CVD process parameters (namely substrate pre-treatment, substrate temperature, reaction pressure, and gas mixture composition), diamond thin films with a variety of grains, namely epitaxial, highly oriented, polycrystalline, and nanocrystalline grains, can be obtained. When diamond forming gas species reach the substrate's surface, adhere to it, and settle quickly into possible equilibrium positions before any structural defects form on the thin film growth front, a single crystalline

IntechOpen

diamond thin film is obtained. On the contrary, when the diamond forming gas species do not quickly settle into stable equilibrium positions upon their arrival at the substrate's surface, nanocrystalline diamond film is obtained.

With the extraordinary and diverse properties that diamond thin films possess, they have a wide range of application potential [3, 4]. However, to realize the complete application potential of diamond thin films, their surface modification is inevitable, especially in surface-sensitive sensing, electro-mechanical systems, and electrochemical applications. Any surface modification process typically modifies a selected diamond thin film's surface by bringing physical, chemical, or biological characteristics different from those found on the surface of the as-deposited diamond film. Methods such as plasma and ion etching, colloidal crystal templating method, selective area CVD, molding, traditional lithography are available to obtain patterned diamond surfaces. Single crystal, as well as polycrystalline diamond surfaces, were exploited to fabricate cantilevers, whiskers, tips, and gratings.

In this chapter, unique surface modifications, namely the direct patterning of sub-wavelength features on the surface of a diamond thin film using a femtosecond laser [5, 6], rapid thermal annealing (RTA) as a means to prepare the surface of a diamond thin film as an efficient direct charge transfer surface-enhanced Raman scattering (SERS) substrate [5, 7], and implantation of $^{14}N^{+}$ ions into the surface and sub-surface regions of a diamond thin film to enhance its surface electrical conductivity [8] will be discussed. The surface of a typical polycrystalline diamond thin film (deposited on Si substrate using CVD) will be considered as the surface to be modified. The discussion will encompass the modification method and post-modification characteristics of modified diamond thin films.

2. Direct surface patterning using femtosecond laser

Irradiation with a femtosecond (fs) laser beam has been touted as a better technique for surface patterning (through ablation) of ultra-hard materials. In this context, the single-crystal diamond surface has been micromachined or patterned by irradiating the surface with high-energy fs-laser pulses. Surfaces of polycrystalline diamond thin films have also been patterned using fs-laser irradiation. In this section, direct surface patterning on polycrystalline diamond thin film using polarization-controlled surface plasmon (SP)-fs laser coupling mechanism will be explained [5, 6]. The polycrystalline diamond film considered here was grown on Si (100) substrate using microwave plasma-enhanced CVD for 6 h at a substrate temperature of 700°C and gas pressure of 25 Torr. Less than 1% of methane (CH_4) in hydrogen (H_2) was used as the reaction gas mixture. Before the thin film deposition, the Si substrate was scratched manually against a thick quartz plate with a synthetic polycrystalline diamond paste (~1 μm) in between. After diamond seeding, the residue was removed by ultrasonically cleaning the substrate with acetone, followed by cleaning with ethanol, then dried in air at room temperature.

The schematic of the experimental setup used for the direct fs-laser patterning on the diamond thin film surface under ambient conditions is shown in **Figure 1**. Ti:Sapphire oscillator-amplifier system operating at a wavelength (λ) of ~800 nm and delivering 110 fs, 1 mJ output energy pulses at a repetition rate of 1 kHz was used to irradiate the considered diamond thin film surface. The fs-laser pulses are linearly polarized. The laser pulse energy was adjusted to 200–500 nJ using neutral density (ND) filters and a combination of half-wave plate and polarizer. The laser beam was focused onto the diamond thin film surface using an objective with a numerical aperture of 0.4, and patterning was performed transversely to the laser propagation direction. The laser spot size on the surface was ~2.4 μm.

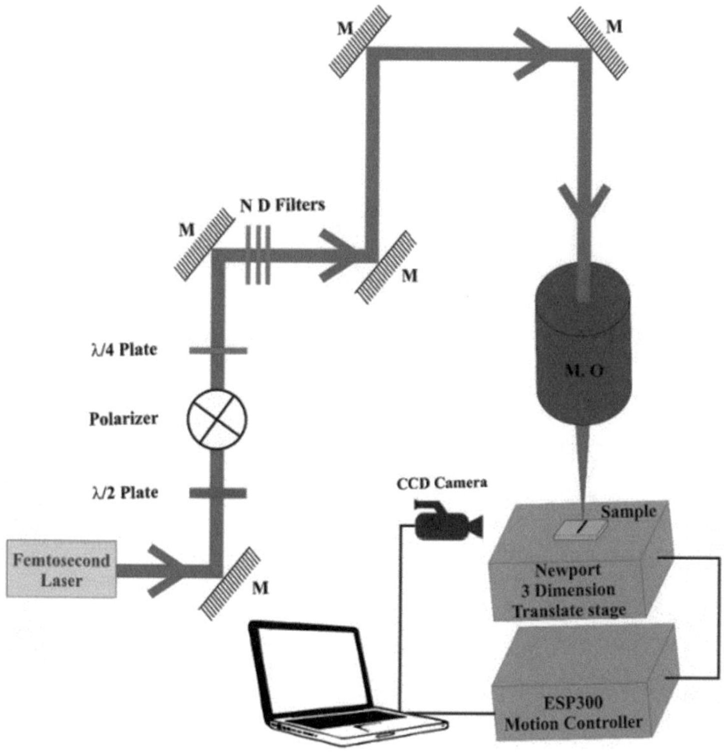

Figure 1.
Schematic of fs-laser direct writing experimental setup [5].

Computer-controlled translational stages were used to translate the sample in x, y, and z directions. A CCD camera with high magnification was used to control the sample surface's focal depth and monitor the patterning process. Transverse patterning geometry (in which the stage is translated perpendicular to the laser beam propagation direction) was used to pattern the surface. Transverse patterning geometry was preferred over longitudinal patterning geometry because it provides much greater flexibility and allows to write patterned structures of arbitrary length.

Plane-view scanning electron micrographs of the as-deposited and fs-laser irradiated diamond thin film surfaces are shown in **Figure 2(a)–(h)**. The surface of the as-deposited thin film (**Figure 2(a)**) is constituted by polycrystalline diamond grains. **Figure 2(b)–(d)** show the low and high magnification plane-view SE micrographs of periodic features induced on the thin film surface with linearly polarized fs laser pulses having a pulse energy 380 nJ. An overview of the fs-laser irradiated region is shown in **Figure 2(b)**. Depending on the laser polarization, morphological and topographical changes are observed in the central area of the irradiated region. Irrespective of the scanning direction, linear periodic ripples perpendicular to the laser polarization and nearly circular ripple features are induced due to linear and circularly polarized light irradiation, respectively. The energy threshold for generating periodic surface features is 200 nJ in the present case. The energy threshold may vary depending on the quality of the diamond thin film considered for surface patterning. The laser-induced periodic surface features have a periodicity of ~180 nm (4.4 times < the free-space λ of the laser and 1.85 times < the λ of the laser in diamond). The width of the fringes is ~80 nm, and the orientations of the fringes are perpendicular to the laser polarization.

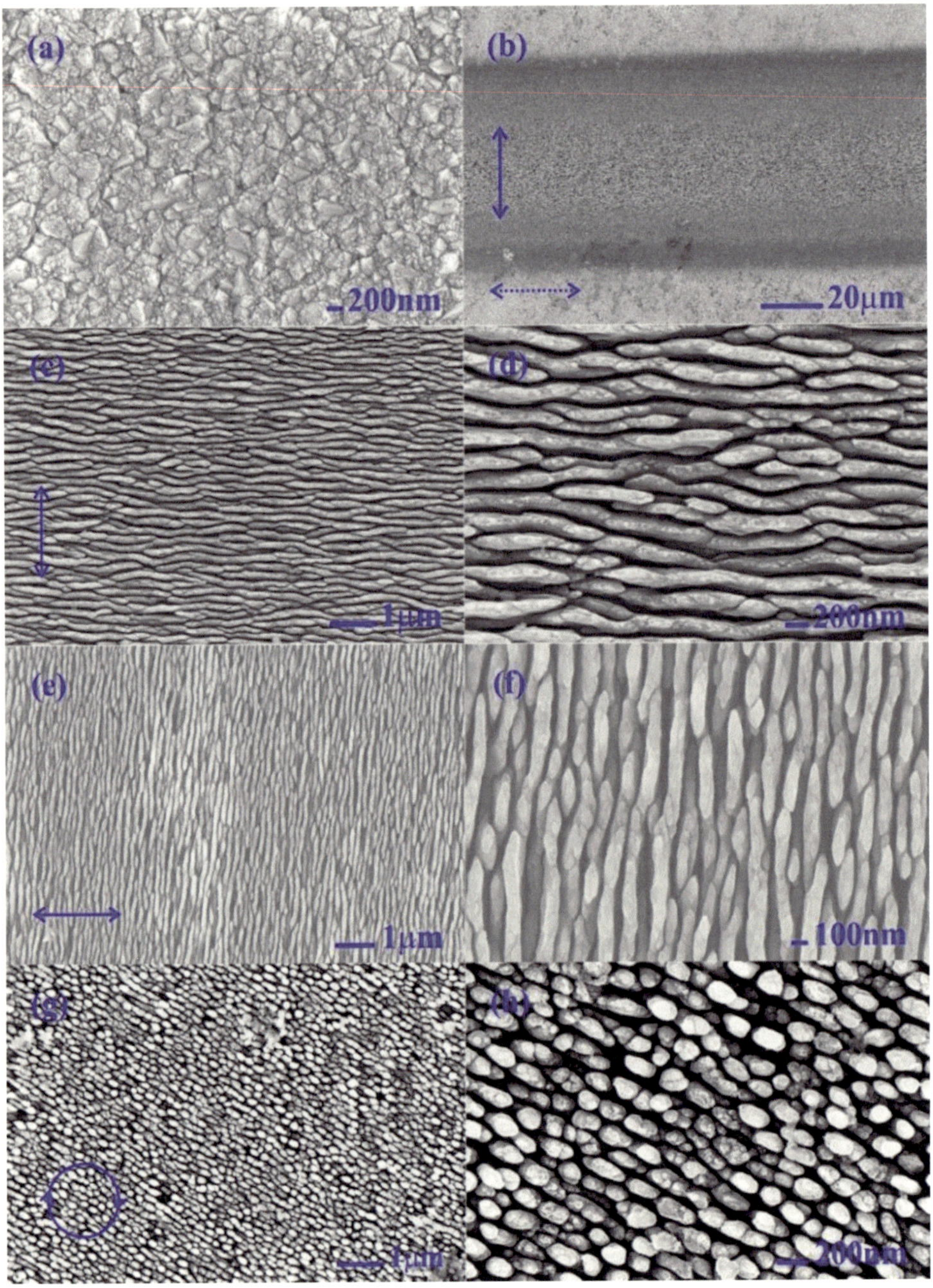

Figure 2.
Plane-view scanning electron micrographs of as-deposited and fs-laser irradiated polycrystalline diamond thin film surface under different experimental conditions: (a) morphology of as-deposited film; (b), (c), (d), (e), and (f) periodic surface features (ripples) obtained using linearly polarized laser pulses having a pulse energy of 380 nJ; (g) and (h) near-circular surface features obtained using circularly polarized laser pulses having a pulse energy of 410 nJ. The polarization and the scanning directions are indicated by the solid and dotted arrows, respectively. In all the above cases, the film surface was moved at ~100 μm/sec w.r.t. to the irradiating fs-laser. (Reprinted with permission from Kuntumalla et al. [6]).

Similar results obtained in parallel and perpendicular scanning (w.r.t. laser polarization) unambiguously show that the scanning direction does not influence the morphology and orientation of the surface features. If the laser polarization direction is flipped by 90° using a half-wave plate in the optical path, the direction of the periodic surface features also flipped by 90°, as shown in **Figure 2(e)** and **(f)**.

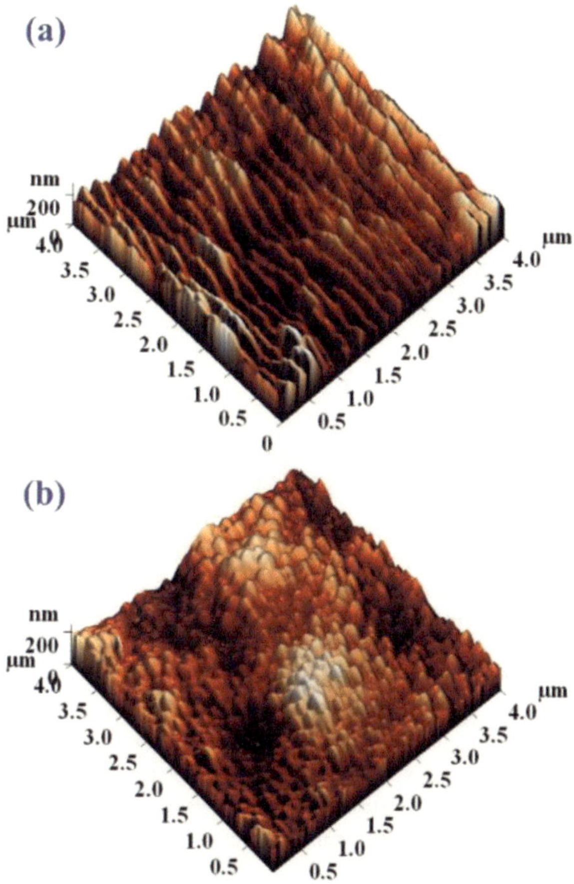

Figure 3.
Surface topography (obtained using atomic force microscopy) of fs-laser irradiated diamond thin film surfaces. (a) Represents features obtained using linear polarized light, while (b) represents features obtained using circularly polarized light. (Reprinted with permission from Kuntumalla et al. [6]).

Irradiation of the thin film surface with circularly polarized fs-laser resulted in nearly circular surface features with lateral sizes in the range ~ 100–200 nm, as shown in **Figure 2(g)** and **(h)**. Raman scattering analysis showed that the fs-laser irradiation had only caused minimal graphitization on the patterned diamond film surface [6]. Similarly, X-ray diffraction analysis showed the crystallinity of the patterned surface was intact. RMS roughness values of laser irradiated surfaces constituting linear (**Figure 3(a)**) and circular (**Figure 2(b)**) surface features were in the range 42–46 nm. This acceptable increase in RMS roughness from the original value of ~30 nm is an important characteristic that is often looked for during surface fabrication methods like the one presented here.

Here, the spatial period (Λ) of the surface ripples formed by the interference of laser with surface plasmons (SPs) is expressed as Eq. (1)

$$\Lambda = \frac{\lambda}{\frac{\lambda}{\lambda_s} \pm sin\theta} \tag{1}$$

where λ and λ_s are the wavelengths of incident laser and SPs, respectively. θ is the incident angle of the fs-laser. λ_s can be obtained from the dispersion Eq. (2)

$$\lambda_s = \lambda \left(\frac{\varepsilon' + \varepsilon_d}{\varepsilon' \varepsilon_d} \right)^{1/2} \tag{2}$$

for metal/dielectric interface. Here, ε' is the imaginary part of permittivity, and ε_d is the dielectric constant of the dielectric material (here air since the experiments are carried out in ambient conditions). Here, the laser is incident normally (i.e., $\theta = 0°$) on the diamond thin film surface and the dielectric constant of air is 1(i.e., $\varepsilon_d = 1$). Therefore, the period can be expressed as Eq. (3)

$$\Lambda = \lambda_s = \lambda \left(1 + \frac{1}{\varepsilon'} \right) \tag{3}$$

Eqs. (1)–(3) are considered here because it is widely accepted that a semiconductor or dielectric (or even metallic) surface such as the diamond thin film surface, when irradiated by ultrashort pulse laser such as fs-laser with damage threshold fluence, will be under highly excited state and behaves like metal surface. When irradiated with fs-laser pulse energy in the range 200–500 nJ, the diamond surface here acts like a metallic liquid carbon fulfilling the metallic state condition, i.e., $\varepsilon' < -1$. In this state, the formation of SPs is facilitated. Once SPs are formed, they undergo interference with the laser light leading to the formation of deep-subwavelength ripples ($\Lambda/\lambda = \sim 0.22$ as experimentally obtained). As the thin film surface is translated, the already formed periodicity favors nonlinear absorption of laser energy at an identical spatial distribution, allowing the periodicity to be maintained. The formation of near-circular periodic surface features (as shown in **Figure 2(g)** and **(h)**) is plausibly due to the axisymmetric distribution of the electric field vector of the circularly polarized laser.

3. RTA to prepare direct charge transfer SERS surface

Surface-enhanced Raman scattering (SERS) based analytical bio-sensing requires stable, durable, and reusable SERS active surfaces. However, the typical SERS surfaces do not have long-term stability, mainly owing to the instability of metal nanostructures deposited on SERS surfaces or due to the intrinsic nature of the SERS surfaces themselves. In this context, a modified diamond thin film surface can be touted as an excellent SERS surface owing to its inherent properties, namely electronic and mechanical properties, chemical stability/inertness, and biocompatibility. Typically SERS activity of diamond surfaces is achieved by depositing suitable metal nanoclusters on their surfaces. The SERS activity in such cases is explained in terms of the electromagnetic enhancement mechanism [5, 7]. In the electromagnetic enhancement mechanism, there is a charge transfer from the Fermi level of the metal nanoclusters to the lowest unoccupied molecular orbital (LUMO) of the adsorbed analyte molecules on the SERS surface. The charge transfer changes the effective polarizability of the adsorbed analyte molecules, resulting in the observed SERS activity.

In this section, the diamond thin film surface modified using rapid thermal annealing (RTA) (i.e., without any metal nanostructures on the diamond thin film surface) will be elucidated as a direct charge transfer resonance-based SERS surface [7]. The diamond thin film considered in the previous section is also considered here for elucidation. The diamond thin film surface was rapidly thermal annealed at 1000°C (heating rate is 40°C/sec, soaking time is 30 sec, and cooling time is 5 min to reach room temperature) under N_2 atmosphere. The heating chamber used for

RTA is shown in **Figure 4**. The chamber consists of a quartz window with halogen lamps as infrared heat sources. The diamond thin film (with its surface facing the halogen lamps) was placed on the bedplate, fixed on an aluminum base plate. Since the heating source was on the top of the film's surface, any changes would start at the diamond thin film surface.

RTA changed the diamond film surface morphology from having sharp features with defined boundaries (**Figure 2(a)**) to a sponge-like surface without any well-defined boundaries. The cross-sectional scanning electron micrograph (**Figure 5**) shows that the sponge-like feature is 100 nm thick, and there is no change in the diamond morphology below the sponge-like layer. **Figure 6(a)** shows a plane-view scanning electron micrograph of the diamond film surface after RTA. It can be observed from **Figure 6(a)** that the annealed diamond film surface has an interconnected wire-like surface morphology, which is entirely different from the original morphology.

Further modifications were not made to the thermally annealed diamond thin film surface (test surface) before testing its SERS activity. Rhodamine6G (R6G) fluorescent dye molecules were used as the probe to understand the SERS activity. R6G molecules exhibit a large Raman scattering cross-section and have been used as probe molecules in SERS-based bio-sensing studies. 10 ml of $5x10^{-5}$ M Rhodamine (R) 6G aqueous solution was dropped onto the test surface and was allowed to

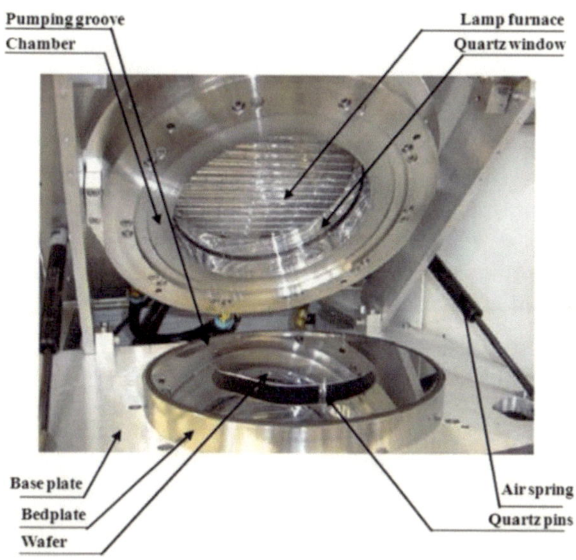

Figure 4.
Rapid thermal annealing chamber [5].

Figure 5.
*Cross-sectional view scanning electron micrograph of thermally annealed diamond thin film shown in **Figure 2(a)** [5].*

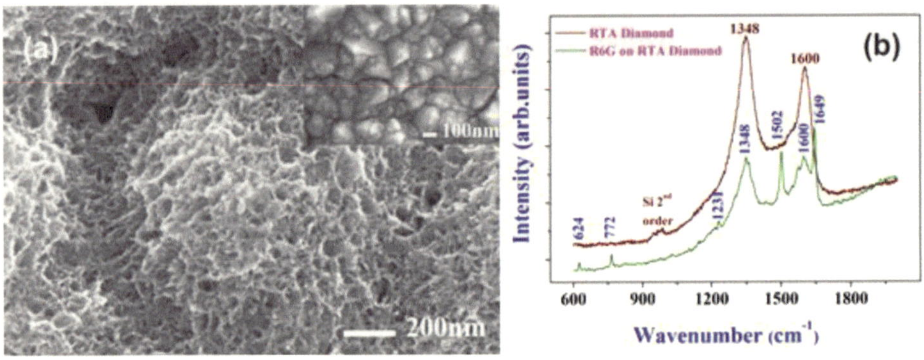

Figure 6.
(a) High magnification plane-view scanning electron micrograph of the thermally annealed diamond thin film surface. Inset: Morphology of as-deposited diamond film (b) Raman spectra obtained from the test surface and R6G ($5x10^{-5}$ M) adsorbed on the test surface. (Reprinted with permission from Kuntumalla et al. [7]).

dry naturally. Room temperature SERS activity of R6G molecules adsorbed on the test surface was recorded using a 514.5 nm laser. When R6G molecules are excited with a visible laser, they show molecular resonance Raman scattering in addition to the SERS effect that comes from the surface plasmon excitation. SERS signals are collected in backscattering geometry in the spectral range 50–2000 cm^{-1} with a spectral resolution of 1 cm^{-1}. **Figure 6(b)** shows Raman spectra obtained from the test surface and R6G adsorbed on the test surface. The Raman spectrum of the test surface shows two broad bands at 1348 and 1600 cm^{-1} correspond to polycrystalline- and nanocrystalline- graphite, respectively. The observation of graphite Raman bands confirms phase change from diamond to graphite due to RTA. SERS spectrum of R6G adsorbed on the test surface shows bands at 624, 772, 1502, and 1649 cm^{-1} corresponding to R6G in addition to the above-mentioned graphite bands.

The observation of strong SERS bands corresponding to very low concentration R6G molecules is a consequence of the charge transfer resonance process because there is no involvement of metal nanoclusters. It is important to note that Raman bands could not be detected when the same concentration of R6G molecules was adsorbed directly on the as-deposited diamond thin film surface. In the case of the as-deposited diamond thin film surface, the energy gap between the Fermi level of diamond surface (111) (−4.97 eV) (or – 5.3 eV for (100) surface) and the LUMO (−3.40 eV) of R6G molecule is about 1.57 eV (or 1.9 eV), which is closer to 2.41 eV, the energy of the incident laser. Hence signal enhancement should take place through the charge transfer resonance process. However, the as-deposited diamond thin film surface did not show SERS activity. In other words, even though the energy of the incident laser is higher than that of the energy gap between the Fermi level of diamond and LUMO of R6G, there is no charge transfer from diamond surface states to R6G molecule. In diamond, the carbon atoms are bonded in sp^3 hybridization (σ-bond). The heavily lopsided configuration of the sp^3 orbital allows a substantial overlap and a strong covalent bond (bonding electrons are tightly bound to the nuclei of the bonding carbon atoms) when the atom combines with the sp^3 orbital of another carbon atom. The thermally annealed diamond thin film surface is constituted by substantial amounts of graphitic-like sp^2 bonded carbon. In sp^2 bonded structure, three of four valence electrons of carbon are covalently bonded (σ-bond) while the fourth is free. Such delocalized electrons participate in π-bond formation between carbon atoms and can traverse between them. In the thermally annealed diamond thin film surface, the amount of sp^2 bonded carbon is considerably larger when compared with the as-deposited diamond thin film surface. Therefore, in the thermally annealed diamond thin film, an enormous number of delocalized electrons are available to interact with the R6G molecule when the laser falls on R6G adsorbed test surface. This results in SERS activity thermally annealed diamond thin film. To further confirm, a typical diamond-like carbon (DLC) thin film surface was also tested (in similar lines to the thermally annealed diamond thin film surface), owing to its constitution by a considerable amount of sp^2 bonded carbon. DLC thin film also showed excellent SERS activity, proving that the thermally annealed diamond thin film surface is an excellent direct charge transfer resonance SERS surface. Additionally, due to the Metal (100 nm of a graphitic layer)-Insulator (diamond)-Semiconductor (Si substrate) (MIS) structure (**Figure 5**), it is easier to develop bio-sensing devices.

4. $^{14}N^+$ ions implantation to prepare an electrically conductive surface

Polycrystalline diamond thin films deposited on doped Si substrates can be made suitable for the fabrication of robust MIS type devices when the near-surface regions are made electrically conductivity. In addition to the RTA presented in the previous section, implantation of N^+ ions into the surface and sub-surface regions of a typical polycrystalline diamond thin film can be another way of achieving MIS structure. The polycrystalline diamond thin film (named F1) used in the previous two sections is again considered for the implantation of N^+ ions. The diamond thin film surface is implanted with 100 keV $^{14}N^+$ ions with fluences of 1E16 and 1E17 ions/cm^2 (the samples are referred to as F2 and F3, respectively). Implantation experiments are carried out at room temperature by maintaining a constant beam current of 10 μA under pressure < 10^{-7} mbar.

XRD patterns of as-deposited and N^+ ions' implanted diamond thin films are shown in **Figure 7**. The reflections corresponding to (111), (220), and (400) crystal planes in diamond are observed at 2θ = ~45.9, ~75.2, and ~ 91.4°, respectively, in all the samples implying that the crystallinity of the diamond thin films after implantation is intact. Moreover, no other reflections corresponding to any different crystalline phase were observed post-implantation, suggesting that the implantation did not convert diamond into any new phase. The diffraction peak broadening post-implantation is negligible, indicating that the implantation has not affected the

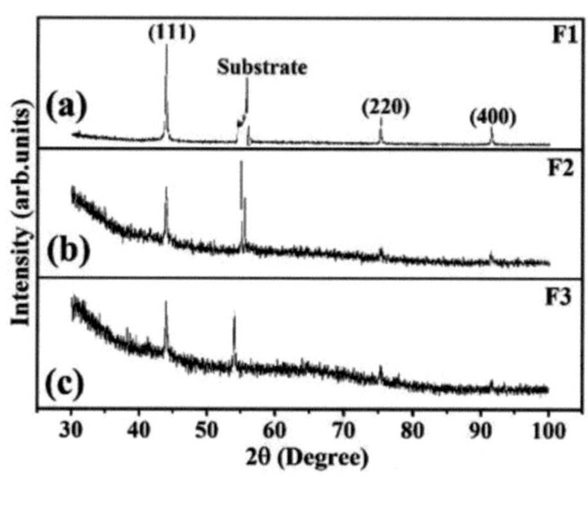

Figure 7.
X-ray diffractograms of (a) F1, (b) F2 and (c) F3. (Reprinted with permission from Bommidi et al. [8]).

grain size in the diamond films. **Figure 8** shows the plane-view scanning electron micrographs, which depict the typical randomly oriented diamond grains of F1 and N^+ ions' implanted diamond films (F2 and F3). However, in the case of F2 and F3, the grain edges appeared to be rounded-off most plausibly due to the selective sputtering effect of the N^+ ion beam on the grain boundary regions, which are primarily graphitic (sp^2 carbon) in nature.

To further understand the phase stability of diamond and the presence of any other phases in small quantities, the Raman spectra of all the samples are recorded and shown in **Figure 9**. F1 exhibited the characteristic zone center phonon in diamond at ~1333 cm^{-1} and graphitic band (G band, stretching mode of planar sp^2 C at the grain boundaries) at ~1500 cm^{-1}, which are typical for diamond films. The spectra of F2 and F3 indicated a change in the bonding environment in the implanted films. The peak fitted spectra are shown in **Figure 9(d–f)**. It is observed that the position, relative intensity, and bandwidth of diamond band, D band (1350 cm^{-1}, breathing mode of planar sp^2 C at the grain boundaries), and G band are different in F1, F2, and F3, indicating differences in phase content in these samples. Diamond, and D and G Raman bands in the case of F2 and F3 are slightly red-shifted compared to that in F1, indicating that implantation has induced in-plane tensile stresses and formation of sp^2 clusters and amorphous carbon in the diamond grain boundaries, respectively. Rutherford backscattering spectra (**Figure 10**) of all the samples are recorded to understand implantation depth. In F2 and F3, both C and N peaks are observed in the corresponding spectra. The width of N peak is ~75 nm which is close to the calculated (Stopping and Range of Ions in Matter (SRIM) software-based) depth of 88 nm for N atoms in diamond.

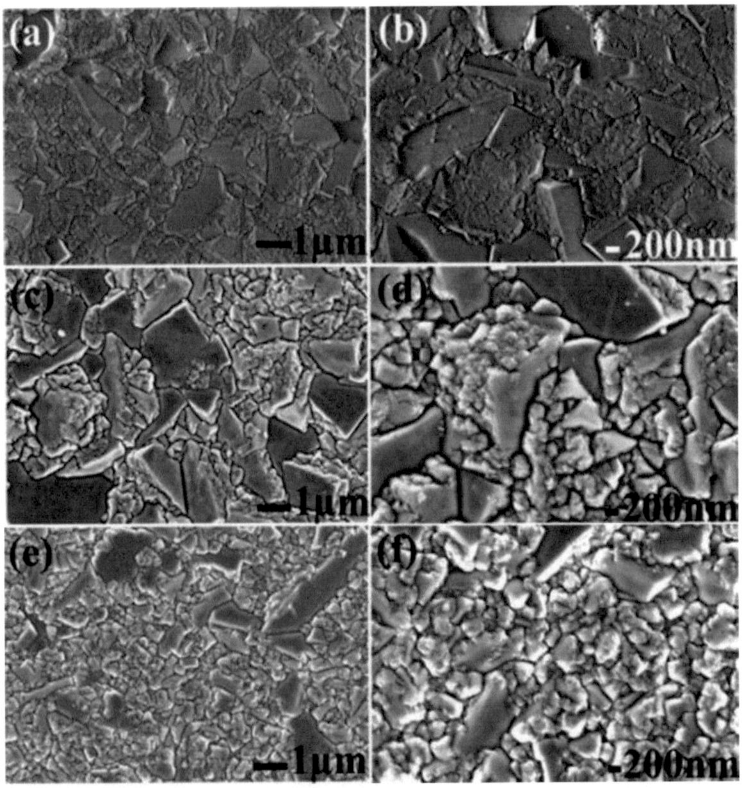

Figure 8.
Plane-view scanning electron micrographs (at different magnifications) of (a & b) F1, (c & d) F2 and (e & f) F3. (Reprinted with permission from Bommidi et al. [8]).

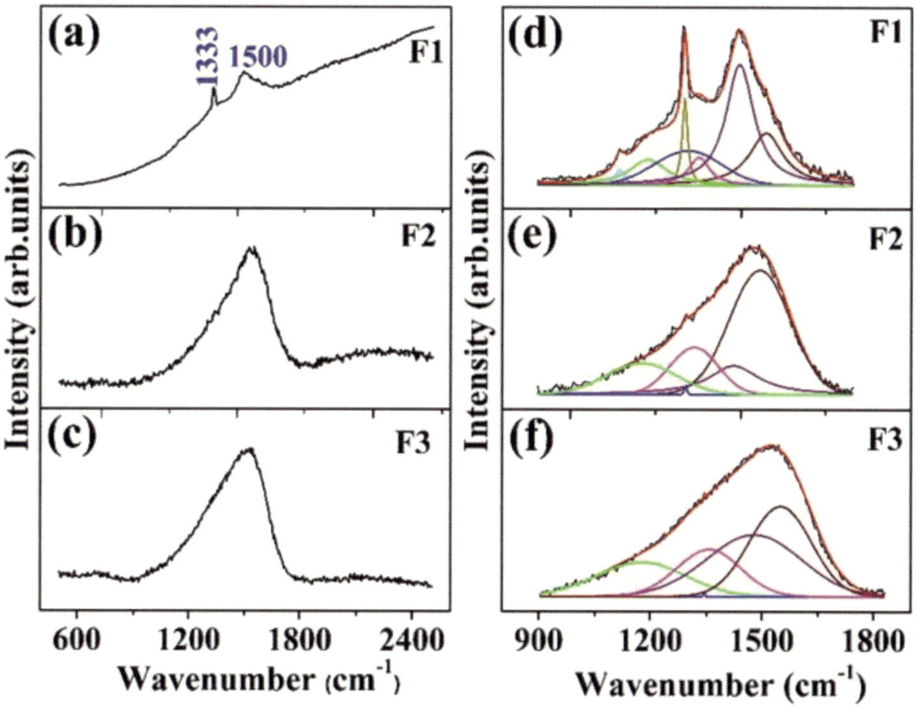

Figure 9.
Raman spectra of (a) F1, (b) F2 and (c) F3 and the corresponding best fitted spectra in (d), (e) and (f), respectively. (Reprinted with permission from Bommidi et al. [8]).

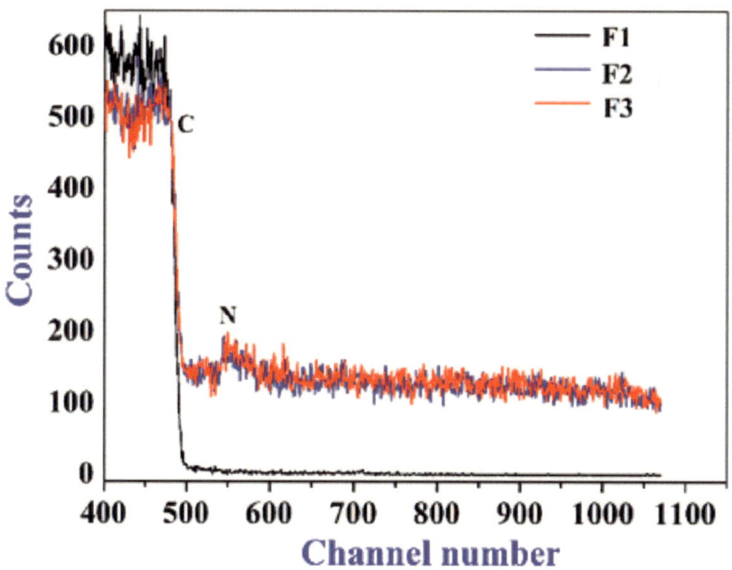

Figure 10.
Rutherford backscattering spectra of F2 and F3 in comparison to F1. (Reprinted with permission from Bommidi et al. [8]).

A local thermal spike (local temperature rise to a great degree) might have taken place in ion-irradiated diamond films [9]. As a consequence, graphitic content might have formed in F2 and F3. The combined effect of induced graphitization along the diamond grain boundaries and N atom incorporation is expected to

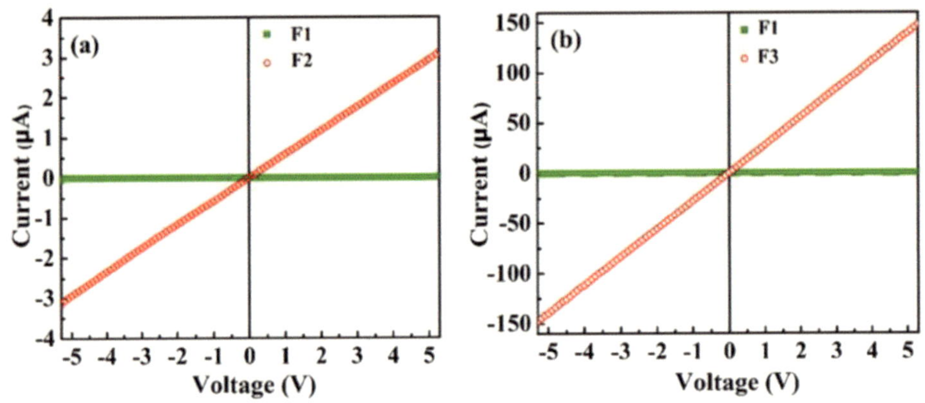

Figure 11.
I-V characteristics of (a) F2 in comparison to F1 and (b) F3 in comparison to F1. (Reprinted with permission from Bommidi et al. [8]).

enhance electrical conductivity significantly. As expected, the I–V characteristics (**Figure 11**) of F2 and F3 in comparison to F1 indicate a considerable increase in the electrical conductivity in them. At 5 V, the current values in F1, F2, and F3 are 4.4 nA, 3 μA, and 140 μA, respectively.

5. Conclusions

In this chapter, three unique surface modifications on a typical polycrystalline diamond thin film have been explained. The first modification has been direct patterning of deep sub-wavelength periodic features on the diamond thin film surface using linearly or circularly polarized fs-laser irradiation. It has also been shown that the surface features are perpendicular to the laser polarization. In other words, it has been demonstrated that laser polarization can be used to control the surface feature formation. The results discussed in Section 2 manifested that the interference of laser with surface plasmons is responsible for forming the surface features without any considerable changes in phase and crystallinity of the diamond thin film surface. Further studies should be taken up to control the shape and size of surface features by controlling different laser parameters such as energy, scan speed, polarization, etc. In Section 3, RTA was used to easily modify the top 100 nm of a polycrystalline diamond thin film surface into a surface with a considerable amount of sp^2 carbon, facilitating an excellent metal-free SERS activity as a consequence of the direct charge transfer resonance process. Moreover, RTA helped in fabricating an MIS-type structure with ease. SERS-based sensing study should be taken up to detect bio-molecules in real-time with an MIS-type biosensor. In Section 4, the N^+ ion beam was used to alter the sp^2 bonded carbon networks and incorporate N atoms into the surface and sub-surface regions in the typical polycrystalline diamond films. Consequently, the electrical conductivity of the diamond thin film was enhanced in several orders compared to the as-deposited film. By controlling the fluence of the ion beam, electrical conductivity can be controlled to a certain depth of the diamond thin film without considerable changes in the morphology and phase of diamond in the film. N^+ ion beam implantation also helped in fabricating an MIS-type structure with ease. Moreover, this method can be useful, especially when large-area diamond thin films have to be modified to make their surfaces electrically conductive.

Author details

Vadali Venkata Satya Siva Srikanth
School of Engineering Sciences and Technology, University of Hyderabad, Hyderabad, India

*Address all correspondence to: vvsssse@uohyd.ac.in

References

[1] Vadali V S S S, Jiang X. Synthesis of diamond films. In: Brillas E, Martínez-Huitle C A, editors. Synthetic Diamond Films: Preparation, Electrochemistry, Characterization and Applications. Hoboken: John Wiley & Sons, Inc.; 2011. p. 21-55. DOI: 10.1002/9781118062364.ch2.

[2] Srikanth V V S S. Review of advances in diamond thin film synthesis. Proceedings of the Institution of Mechanical Engineers, Part C: Journal of Mechanical Engineering Science. 2011; 226:303-318. DOI: 10.1177/0954406211422788.

[3] Kratochvilova I. Polycrystalline diamond thin films for advanced applications. In: Silva AMT, Carabineiro SAC, editors. Advances in Carbon Nanostructures. London: IntechOpen; 2016. p. 161-173. DOI: 10.5772/64701.

[4] Brillas E, Martínez-Huitle C A, editors. Synthetic Diamond Films: Preparation, Electrochemistry, Characterization and Applications. Hoboken: John Wiley & Sons, Inc.; 2011. 632 p. DOI: 10.1002/9781118062364.

[5] Kuntumalla M K. Modification and enhanced Raman activity of nano-diamond and β-SiC thin film surfaces and Kelvin force microscopy and surface phonons of diamond/β-SiC nanocomposite thin films [thesis]. Hyderabad: University of Hyderabad; 2016.

[6] Kuntumalla M K, Rajamudili K, Desai N R, Srikanth V V S S. Polarization controlled deep sub-wavelength periodic features written by femtosecond laser on nanodiamond thin film surface. Applied Physics Letters. 2014; 104:161607. DOI: 10.1063/1.4873139.

[7] Kuntumalla M K, Srikanth V V S S, Ravulapalli S, Gangadharini U, Ojha H, Desai N R, Bansal C. SERS activity of Ag decorated nanodiamond and nano-β-SiC, diamond-like-carbon and thermally annealed diamond thin film surfaces. Physical Chemistry Chemical Physics. 2015; 17:21331-21336. DOI: 10.1039/c4cp05236f.

[8] Bommidi V. A study on uniquely modified diamond, diamond/β-SiC nanocomposite and graphene nitride (g-C_3N_4) surfaces [thesis]. Hyderabad: University of Hyderabad; 2019.

[9] Bommidi V, Kandasami A, Srikanth V V S S. N^+ ion beam irradiation as a strategy to enhance the electrical conductivity of polycrystalline diamond thin films. Materials Letters. 2019; 241:172-175. DOI: 10.1016/j.matlet.2019.01.085.

Section 4

Mechanical

Chapter 4

Laser Treatment CVD Diamond Coated Punch for Ultra-Fine Piercing of Metallic Sheets

Tatsuhiko Aizawa, Tadahiko Inonara, Tomoaki Yoshino, Tomomi Shiratori and Yohei Suzuki

Abstract

CVD-diamond coated special tools have been widely utilized to prolong their tool life in practical production lines. WC (Co) punch for fine piercing of metallic sheets required for high wear-toughness to be free from chipping and damages and for high product quality to punch out the holes with sufficient dimensional accuracy. The laser trimming process was developed to reduce the surface roughness of diamond coating down to submicron level and to adjust its diamond layer dimensions with a sharp punch edge for accurate piercing. The pulsed laser irradiation was employed to demonstrate that micro-groove was accurately formed into the diamond coating. Less deterioration in the worked diamond film by this laser treatment was proved by the Raman spectroscopy. The femtosecond laser trimming was proposed to sharpen the punch edge down to 2 μm and to form the nano-textured punch side surfaces with the LIPSS (Laser Induced Periodic Surface Structuring)-period of 300 nm. Fine piercing experiments were performed to demonstrate that punch life was significantly extended to continuous punching in more than 10,000 shots and that mirror-shining hole surfaces were attained in every shot by regularly coining the nanotextures. The sharp punch edge with homogeneous edge profile was responsible for reduction of the induced damages into work sheet by piercing. The punch life was extended by the ejection mechanism of debris particles through the nanotextures on the punch side surface. The present laser treatment was useful in trimming and nanostructuring the complex-shaped punch edge for industrial application.

Keywords: CVD diamond coated WC (Co) tools, laser adjustment, laser trimming, punch edge sharpening, punch nano-structuring, ultrafine piercing, AISI316L sheets, debris particle ejection

1. Introduction

The CVD (Chemical Vapor Deposition)-diamond coating as well as the PCD (Poly-Crystalline Diamond)-chip have been widely utilized as a protective layer of special tools to prolong their life [1]. In addition to their application to cutting tools, the sintered diamond dies were employed as a tool in the metal forming [2]. In particular, the CVD-diamond coated tools become a standard procedure to cut the CFRP (Carbon Fiber reinforced plastic) members [3], to make deep drawing of stainless steel sheets to cups [4] and to precisely punch out the high strength copper alloy plates [5]. During those manufacturing processes, the diamond coating is usually damaged and chipped so that the tool substrates have to be recycled by removing away or ashing the used coatings before being wasted as an industrial dust.

Let us remember that WC (Co) and silicon were only utilized as a substrate for hot-filament CVD processes to have thick, uniform diamond layer. In particular, the cobalt content in WC (Co) substrate is optimally selected for proper nucleation and growth of diamond coating in practice. In order to improve the material efficiency in the coming circular economy, these used DLC- and diamond-coated cutting tools and forming dies must be recycled to reuse the original WC (Co) substrate for higher cost-competitiveness. In this circulation of WC (Co) substrates [6], the perfect removal or ashing of diamond film and metallic buffer layers with minimum damage to tool geometry is an essential process to reuse the WC (Co) as illustrated in **Figure 1**. RF (Radio-Frequency) – DC (Direct Current) plasma ashing process was developed to make perfect removal of the used DLC coatings without significant damages to substrates and to reuse the as-ashed WC (Co) substrate for recoating [7–12]. As pointed in [13], the oxygen ion density in those ashing processes must be intensified enough to remove the used CVD-diamond coating with sufficiently high removal rate. This high density plasma ashing method has proper capacity to remove the diamond films even on the rake surfaces of cutting tool blades with less blade edge loss than 1 μm [14–16]. Hence, as a challenging issue in **Figure 1** for circulation economy of WC (Co) tool substrates, the CVD-diamond coated tools must be shaped to have high capability for cutting and shearing in practical operations and to improve the total WC (Co) efficiency.

As-coated diamond film has a rough surface due to its three dimensional crystalline growth; the maximum surface roughness must be reduced down to the tolerance in the industrial applications, less than 0.5 μm. Its geometrical profile is never adaptive to the precise stamping tools; its surfaces must be trimmed to have accurate dimensions as a tool for fine cutting, shearing and piercing within the deviation of 1 μm. In addition, the diamond-coated tools and dies must have sharp edges enough to preserve the highly burnished surfaces of products. Furthermore, their lives must be elongated by reducing the adhesion of work material debris.

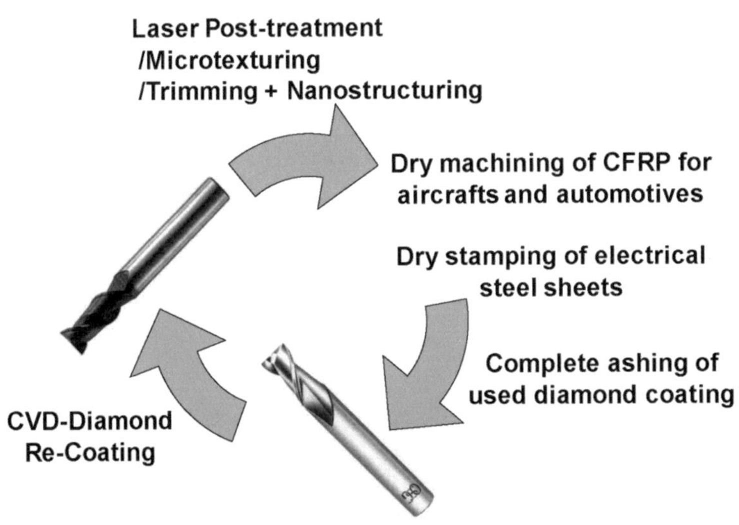

Figure 1.
A circular economy of WC (Co) tools by the precise treatments of diamond coatings.

In the present chapter, among the laser processings [17, 18], the laser treatment of CVD-diamond coated tools is proposed to geometrically adjust their diamond coating profile, to trim their surfaces and sharpen their edges and to form the nanostructured micro-grooves for in situ ejection of debris particles during piercing process. Since the first world-wide notice on the importance of wear debris [19], a role of debris particles on the tribological performance in metal forming has been studied both in academic and industries. In particular, fine debris fragments induced the fretting wear in metal forming; how to eject those debris particles still becomes an issue of nuisance [20]. Hence, this laser nanostructuring technique is a powerful approach to prolong the fine piercing punch life and to control the piercing process for both ductile and brittle work sheets.

Furthermore, this laser nanostructuring provides a method to design and fabricate the engineered surfaces to mechanical elements such as the channel, the orifice and the heat sink. If their inner and outer surfaces have an appropriate nanostructure, these structured surfaces are expected to work as an engineered surface with higher wettability, more hydrophobicity and larger overall heat penetration. In the case of the orifice, the leak flow of gasoline through the orifice walls is minimized by their hydrophobicity of nanotextured surface [21, 22]. A nanotextured heat sink has a capacity to significantly improve the heat transfer coefficient, especially the boiling water heat transfer capacity [23, 24]. In particular, the AISI316L orifice plate is a target for the present fine piercing by the nanostructured surface to improve its surface property control.

A CVD-diamond coated WC (Co) specimen is first employed to demonstrate that the pulsed laser adjustment is effective to shape the diamond coated tools and to make microtexturing to diamond films without significant damage to the diamond structure. Next, the femtosecond laser micro-machining is employed to trim a diamond-coated piercing punch. This laser trimming enables to reduce the roughness of as-coated diamond films and to sharpen the punch edge width down to 2 μm. Furthermore, the nanotextures with the period of 300 nm are formed from the edge to the specified length on the punch side surfaces, simultaneously with trimming. Finally, the micro-stamping system is utilized to describe the piercing behavior of AISI316L austenitic stainless steel and amorphous steel sheets with the use of this laser trimmed punch. When using the WC (Co) punch with the sharpened edge, its piercing of AISI316L sheets partially induced a fractured hole surface; the burnished surface area ratio was limited by 70 to 80% of their whole pierced hole surface [25]. In addition to fine piercing performance with full burnished area ratio, the nanotextures on the punch side surface is concurrently transcribed onto the AISI316L holes together with this trimming. Due to this imprinting of the laser-trimmed punch surface with nanostructures, a mirror-shining hole surface is fabricated also to have a periodic nanotexture. An amorphous electrical steel sheet is also employed to investigate the piercing performance of brittle work materials [26]. The sharp edge profile and the nanostructured punch side surface have influence to reduce the damaged width and to improve the product quality. SEM (Scanning Electron Microscopy) and WIS (White Interference Spectroscopy) analyses are utilized to describe this formation of nanostructures into diamond-coated punch and their duplication onto the product surface.

2. Laser treatment of CVD-diamond coated tools

Two types of laser treatment system are proposed to make geometric adjustment of CVD diamond coated punch with the use of pulsed laser irradiation and to trim the punch edge and make nanostructuring onto the side surface of punch with the use of femtosecond lasers. These two laser treatments are illustrated in **Figure 2**.

2.1 Geometric adjustment by excimer laser irradiation

As-CVD-coated diamond film with the thickness of 20 μm has a surface roughness of 3 to 5 μm. A sizing treatment [27, 28] process is necessary to reduce the surface roughness within tolerance for tooling in metal forming and to fit the surface profile of punch into tailored geometry for fine piercing and embossing. The pulsed laser irradiation process is employed to remove the unnecessary surface parts of CVD-diamond film through a series of shots in order that the whole surface profile should be fine enough to satisfy the designed CAD (Computer Aided Design) data of tools.

Figure 2a illustrates an experimental set-up for this sizing treatment. As-coated punch was fixed into a jig, which was located on the X-Y stage. With the use of this stage and Z-positioning controller, the work area on the punch surface was located

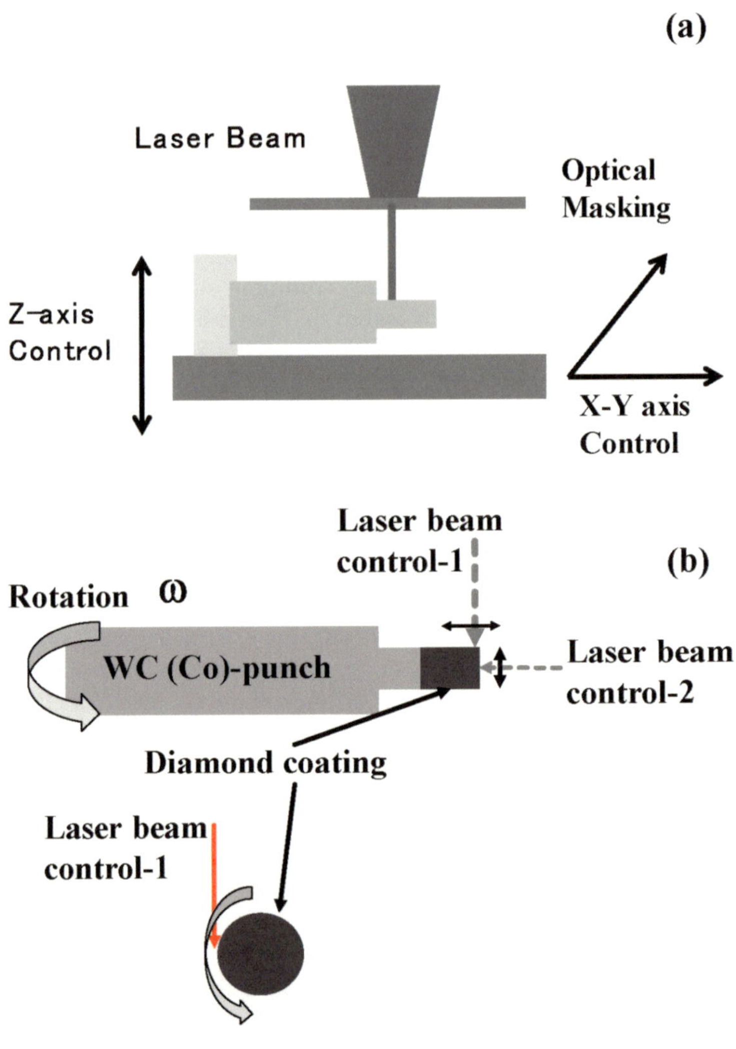

Figure 2.
Two types of laser treatments for the diamond coated tools for fine and ultrafine piercing. (a) Geometric adjustment by pulsed laser irradiation, and (b) trimming by femtosecond laser machining.

for laser irradiation. After the pulsed irradiation, this work area was relocated for next irradiation. The number of pulses was directly controlled to correspond to each feeding depth for removal of diamond coating. A laser spot area was also controlled by optical masking; e.g., one segment on the mask became a transparent window for the pulsed laser beam to irradiate this segmented surface area of work. In this process, the removed thickness of CVD coating by single shot via laser abrasion was optimally determined to be around 0.1 μm by controlling the power profile of laser beams. The original laser beam was modified by the optical masking technique to focus only onto the segment of 250 μm x 125 μm. In the following experiments, the diamond coated punch was controlled to move stepwise in the X axis by 250 μm to form a rectangular micro-groove onto the diamond coating with the width of 125 μm.

2.2 Trimming and nanotexturing by femtosecond laser irradiation

The femtosecond laser micromachining is suitable to trimming and nanotexturing the CVD-diamond coated WC (Co) punches [29–31] for ultrafine piercing of metallic sheets [29–32]. **Figure 2b** depicts the standard setup for this laser trimming and nanotexturing. The side surface was first trimmed by utilizing the laser beam control-1. The punch head was secondly processed by using the control-2. The laser beam was moved from the center to the end of the punch head. The fluence was also held constant at 0.265 J/cm^2. During this two-step procedure in the experiment, the end of the punch was held in a jig to be rotated with a constant velocity by ω = 7.2 degrees/s. The galvanometer was utilized to distribute the laser beam as tailored by CAM (Computer Aided Machining) data for the trimming operation. The capacity of present femtosecond laser machining system is stated in the following. The wavelength of the femtosecond laser was 515 nm, the pulse width was 200 fs, and the pulse repetition rate was 400 kHz. The maximum average power was 40 W, and the maximum pulse energy was 50 μJ. The working stage was 300 mm × 300 mm. A work material with sized 280 mm × 150 mm was placed on the X- and Y-axes controlled stage in **Figure 2b**. The single-shot power was estimated to be 0.25 GW. High-powered irradiation of 200 fs was used to drive the well-defined ablation into the targeted materials.

The laser nanostructuring method stands on the LIPSS (Laser Induced Periodic Surface Structuring) performance [33, 34]. The directional nanotexture was in situ formed together with the laser trimming during the femtosecond laser machining process with a skew angle against the beam scanning direction. In the following experiments, the fluence was constant with 0.6 J/cm^2. The laser machining track overlapped the working range 20 times by rotating the work. The LIPSS-ripple period was controllable by the laser fluence, pulse width, and so forth for femtosecond laser nanotexturing. In fact, LIPSS using high and low spatial frequencies with very different periods can be produced via the same laser setup, depending on the process conditions. This LIPSS-ripple period was estimated to be 250 nm in the present trimming conditions. To be discussed later, this LIPSS period as well as the nanostructuring alignment are controllable by the laser processing conditions during the optical polarization and transformation processes.

3. Micro-grooving of diamond-coated tools by pulsed laser irradiation

An excimer laser machining system (LIPS-Works, Co., Ltd.) was employed to make pulse laser irradiation of CVD-diamond coated rectangular punch. A micro-groove was formed onto the diamond coating to describe the dimensional accuracy

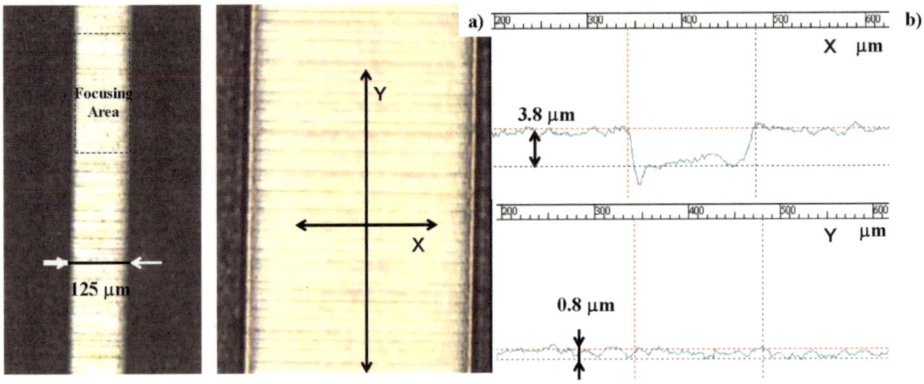

Figure 3.
Micro-grooved diamond coating by the pulsed laser machining. (a) a micro-groove with the width of 125 mm with low and high magnifications, and (b) its surface depth profiles along the X- and Y-axes.

in sizing treatment of diamond coated punch. Raman spectroscopy was employed to characterize the diamond film before and after this micro-grooving process.

3.1 Micro-grooving of diamond-coated tool substrate

The excimer laser was employed to make micro-grooving onto the CVD-diamond coating. In each pulsed laser irradiation, the area of 250 μm x 125 μm x 0.1 μm was stepwise removed by a single shot. After multi-shot irradiation, the punch was relocated to move in the X axis by 250 μm to continue this laser ablation process till the end of punch width. The number of laser pulse shots is constant by 50.

Figure 3a depicts the laser-machined track of CVD-diamond coating. Without damages to the un-irradiated surface, only a single micro-groove with the width of 125 μm is accurately cut-in by the present laser machining. This linear removal of coating takes place only with positioning control of specimens without any change in the laser irradiation conditions. The feeding depth is controlled by the number of shots independently from the above spatial control. This results in precise profiling of CVD-coated tool surface geometry in the suitable manner to tooling design. The laser-microscope (Laser-tech, SD 100; Tokyo, Japan) was utilized to measure the surface roughness distributions both in the longitudinal and the lateral directions of linear track. **Figure 3b** depicts the micro-groove surface depth profiles in the X- and Y-axes, respectively. Its average depth is 3.8 μm, and a deep valley is seen at either edge of micro-groove. This might be because the laser energy profile is intensified at the edge of masking window. The maximum roughness in the longitudinal direction at the bottom of micro-groove is only 0.8 μm. This proves that this sizing process by pulsed laser irradiation accurately adjusts the punch edge profile as demanded by the engineering CAD.

4. Raman spectroscopy on the laser treatment effect on diamond

Raman spectroscopy (Renishaw, Co., Ltd.) was utilized to characterize the effect of the laser adjustment to the microstructure of CVD diamond. **Figure 4** compares the Raman spectra and their deconvoluted profiles before and after laser adjustment. As-coated diamond is characterized by the graphite disordered D peak at

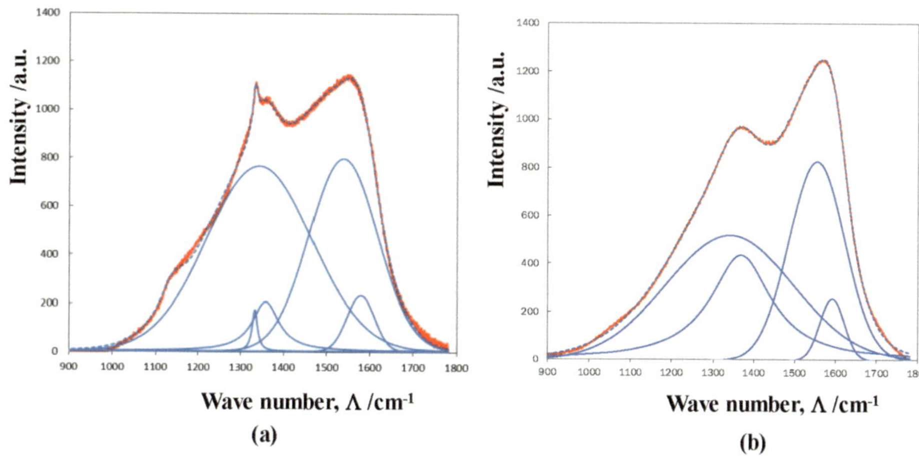

Figure 4.
Comparison of Raman spectra before and after the pulsed laser post-treatment. (a) Raman spectra before treatment, and (b) Raman spectra after treatment.

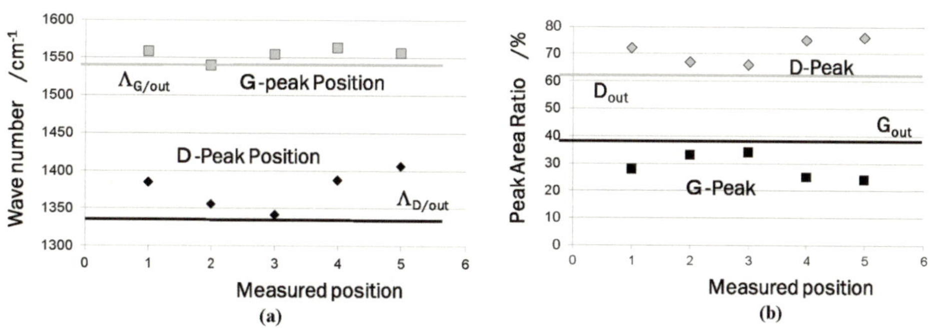

Figure 5.
Characterization on the laser-processed diamond coatings with comparison to unprocessed film. (a) Distribution of Λ_G and Λ_D on the micro-grooved surface with comparison to the average Λ_G and Λ_D outside of the microgrooves, and (b) distribution of peak area ratio for G- and D-peaks on the micro-grooved surface with comparison to the average peak area ratios outside of the microgrooves.

1340 cm^{-1}, the crystalline G peak at 1580 cm^{-1} and sp3 diamond peak at 1320 cm^{-1}. This reveals that CVD-coated diamond film consists of the nano-structured matrix of sp3 – sp2 binding-state carbon and the sp3-rich surface structure. After pulsed laser adjustment, this surface diamond D-peak disappears in **Figure 4b**. The same broad graphitic D and G peaks are deconvoluted from the measured spectra in **Figure 4a** and **b**. Remember that amorphous carbon films are also characterized by these D and G peak pair [10] and that carbon dusts are only detected by low intensity Raman spectra with much broadness [11]. Although the near-surface of diamond coating is affected by laser irradiation, its depth might be characterized by the diamond D-peak as well as these graphitic D and G peak pair [35]. No essential deterioration occurs in this sizing process of the diamond coating. The surface layer with characteristic sp3 nanostructure is only ablazed during irradiation.

The quality profile of processed diamond coating is investigated by analyzing these graphitic D- and G-peak distributions in the longitudinal direction of micro-groove. **Figure 5a** depicts the Raman shift distributions; the graphitic D-peak Raman shift increases to the higher wave number from the center to both ends. As shown in **Figure 5b**, the graphitic D-peak area ratio is nearly constant and higher than 65%. In particular, the measured D-peak area ratio at the micro-groove bottom is always higher than that on the original diamond film. This also proves that the surface of sized diamond coating by the pulse laser irradiation has no significant deviation in quality of coating materials.

5. Femtosecond laser trimming of diamond-coated tools

Femtosecond laser irradiation system was utilized to trim the as-coated diamond film surfaces, to sharpen the punch edge and to make nanostructuring on the side surface of cylindrical punch. SEM with various magnifications was used to make microstructure analysis on these nanostructures. The white light interferometry was also employed to characterize them.

5.1 Trimming and nanostructuring

A CVD-diamond coated WC (Co) punch with the diameter of 2.00 mm was prepared for the present laser trimming and nanostructuring. As shown in **Figure 6a**, the original head and side surfaces of as-coated diamond film are rough by the polycrystalline diamond growth during CVD; e.g., its maximum surface roughness reaches to 5 μm. The punch edge curvature also becomes dull as an intersection of rough head and side surfaces.

Figure 6b shows the laser-trimmed punch profile after surface cleaning. The maximum roughness of punch head is reduced down to 0.5 μm on the measured surface profile. Both head and side surfaces are laser-trimmed so that the punch edge is considered to be sharpened as an intersection of two surfaces.

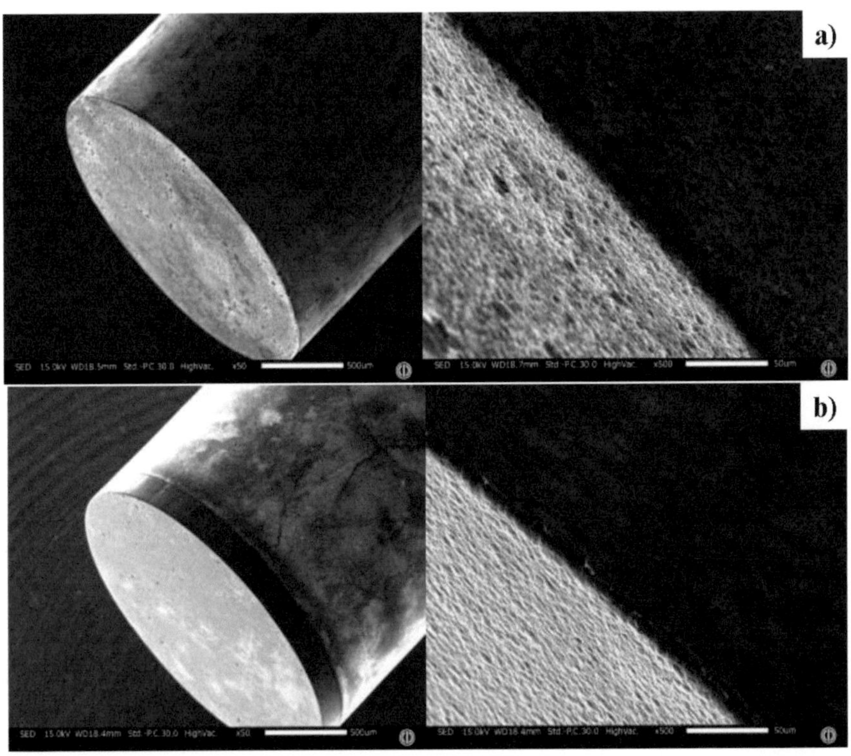

Figure 6.
Comparison of SEM image before and after laser treatment. (a) As-coated head and side surfaces with a dull edge, and (b) laser-trimmed head and side surfaces with a sharpened edge.

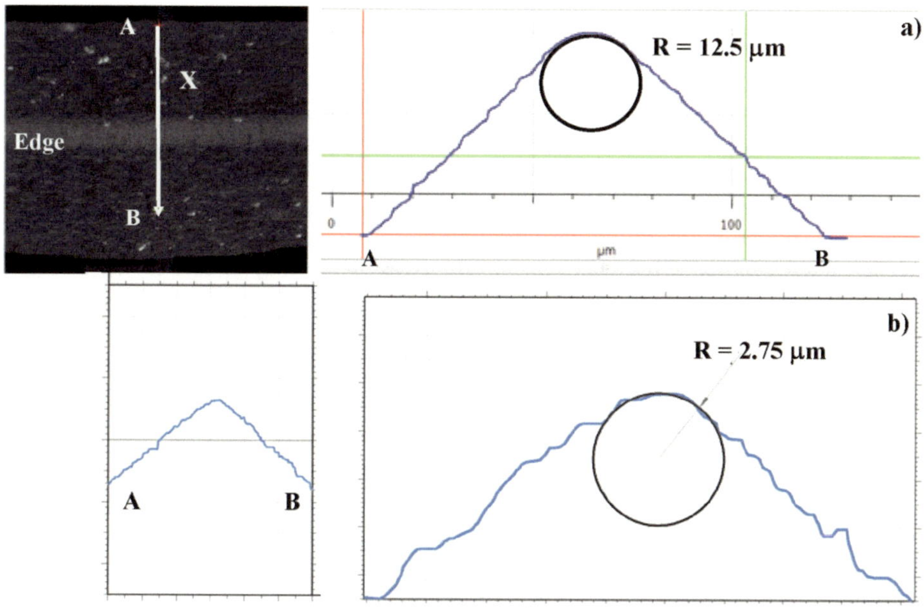

Figure 7.
Comparison of punch edge curvature before and after laser treatment. (a) before laser trimming, and (b) after laser trimming.

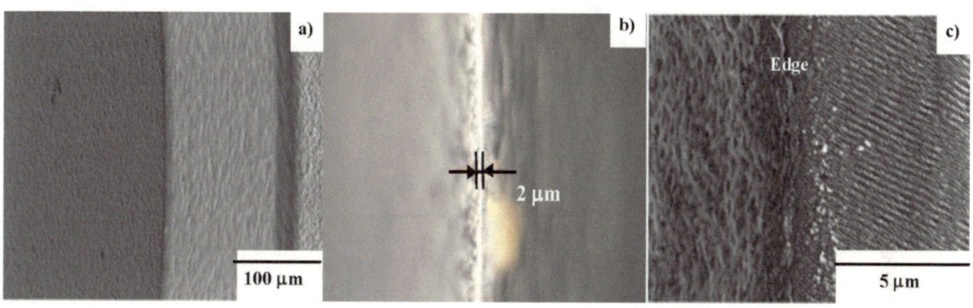

Figure 8.
SEM and LM images on the laser treated side surface from the punch edge. (a) SEM image on the laser-trimmed head and side surfaces of punch in low magnification, (b) LM image around the punch edge, and (c) SEM image across the punch edge in high magnification. A nanostructure was formed to have a regular alignment with its period of 300 nm.

Figure 7 compares the punch edge curvature radius before and after laser trimming. As-coated edge curvature radius (R) is 12.5 μm; this large R is reduced down to 2.75 μm by this trimming. This improvement proves that femtosecond laser trimming is effective to sharpen the punch edge width down to 2 μm.

5.2 Controllability of nanostructures

The femtosecond laser trimming process accompanies with nano-structuring on the trimmed surfaces. When laser-trimming the diamond coating, LIPSS takes place to form the intrinsic nano-textures with the LIPSS-period to the optical interaction between the laser beam scattered by the rough diamond surface and the incident laser beam. SEM was utilized to describe this simultaneous nano-structuring with trimming process in the above. **Figure 8a** shows the SEM image of head and side surfaces of punch. The punch edge width (W_E) is also measured by LM (Laser

Microscopy) in **Figure 8b**. W_E = 2 μm just in correspondence to R = 2.75 μm in **Figure 7b**. **Figure 8c** depicts the change of microstructure from the punch head to its side surface across its edge. The trimmed head surface changes to the nanostructured surface just across the edge. This nanostructure consists of the regularly aligned nano-grooves with the LIPSS period of 300 nm. This measured pitch is corresponding to the estimated LIPSS period of 250 nm when using the above femtosecond laser processing conditions.

White light interferometry (WLI) was utilized as a nondestructive evaluation method for diagnosis of nanotextures on the trimmed punch surface profile. This WLI is usually utilized to measure the geometric angulation of polished and buffed die surfaces with relatively little curvature radius. In this measurement, a trimmed diamond-coated punch with the diameter of 2.00 mm has a curvature with influence on the interferometric measurement on the spatial period of nanotextured ripples on the trimmed surface. **Figure 9a** shows this local surface profile in this X-axis or in the lateral direction of punch surface, which was analyzed by the algorism of DEAP (Detection of Envelope and Absolute Phase) [36]. This profile gradually deviates from the center line; the nanotextures are formed on the trimmed punch surface with the skewed angle in the axial direction. The measured spatial period of nanotextures (Λ_{punch}) is 900 nm, and, their average height reaches to 300 nm.

Let us consider the difference in the spatial period of nanostructures, between the measured Λ_{punch} of 900 nm in **Figure 9a** and the LIPSS-period of 300 nm in **Figure 8c**. In the detection of fine spatial peaks in the large area with the curvature by WLI, the neighboring peaks to a main peak are easily enveloped into a single signal by the DEAP algorism. Then, the WLI-measured period becomes three times more than the actual ripple period of 300 nm. In other words, the curvature effect to the measured profile of peaks cannot be sufficiently eliminated in the present measurement.

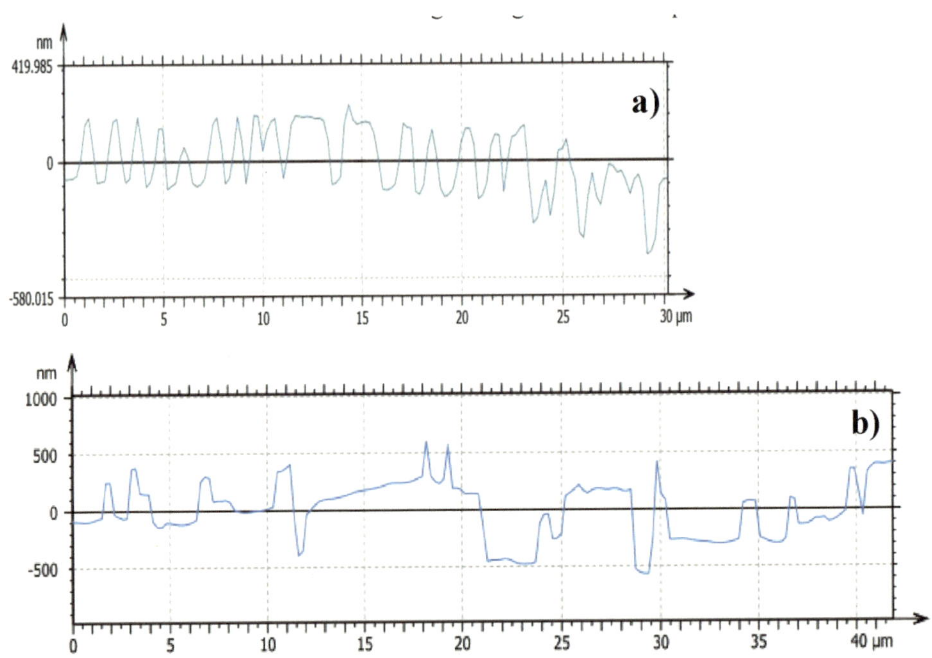

Figure 9.
White light interferometry of the nano-structured diamond-coated punch as well as the imprinted nanotextures onto the hole surface together with the piercing process of AISI316L sheets. (a) Nano-structure on the diamond coating, and (b) imprinted nanotextures on the pierced hole surface.

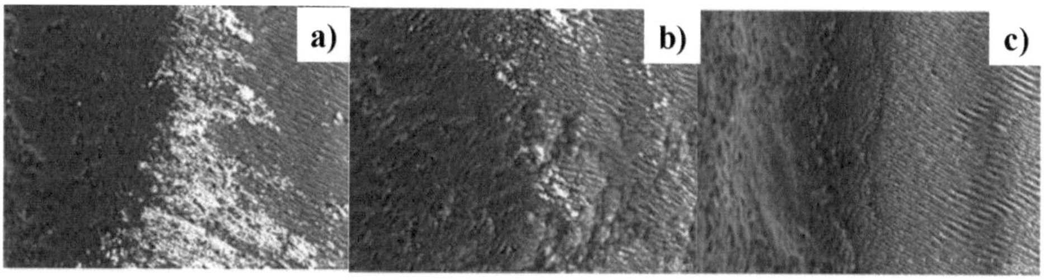

Figure 10.
Effect of the threading depth (d) in the laser trimming process on the nanostructures formed on the punch side surface. (a) d = 1.8 μm, (b) d = 2.4 μm, and (c) d = 3.6 μm.

In the present trimming of the diamond coating on the cylindrical punch, the scattering laser on the coating is skewed by the local curvature on the trimmed diamond surface so that every nanotextured ripple is formed in the axial direction with a skewed angle. This nanotexturing process with trimming the cylindrical diamond coating is mainly governed by this local curvature of cylindrical punch as well as the original roughness of diamond coating.

As pointed out in [33, 34], the laser processing parameters also have influence on this LIPSS or nanotexturing onto the laser-trimmed surface. Among them, the laser pulse width has a direct influence on the ripple-period while the fluence of laser beam affects the LIPSS-profile more than the LIPSS-period. In particular, the depth of nanostructure is incrementally increased by increasing the number of pulses or by increasing the fluence. This laser beam fluence was varied to investigate the effect of irradiation fluence on the depth of nanostructures. **Figure 10** compares the SEM images on the nanotextures on each trimmed surface among three punches. Although the peak-to-valley ratio of nanotextures increases with d, the unidirectional formation of nanotextures with a skewed angle is common to three punches. The measured LIPSS-period is also common to three cases; e.g., Λ_{punch} = 300 nm. The similar nanotexturing profile are simultaneously machined onto the diamond coating with trimming; its depth is mainly determined by the fluence.

6. Ultra-fine piercing of stainless steel sheets by laser-treated punch

AISI316L austenitic stainless steel sheets with the thickness of 0.2 mm were utilized for fine piercing experiments with the use of the treated CVD-coated WC (Co) punch. SEM and white light interferometry were also employed to characterize the quality of pierced products.

6.1 Fine piercing system

Figure 11a depicts a piercing experimental set up where the stroke is controllable in every 1 μm. The relationship between the piercing load and the punch stroke is monitored during this piercing process. The laser trimmed diamond-coated WC (Co) punch with a diameter of 2.000 mm was fixed into an upper die in the cassette die-set for the piercing experiment. The WC (Co) core die with an inner diameter of 2.008 mm was also placed into the lower die. The load cell was embedded into the lower die set to monitor the applied load in every stoke. As illustrated in **Figure 11b**, the narrow clearance between the punch and die is controlled by the nano-metric PZT X-Y stage to preserve the coaxial position of punch to die.

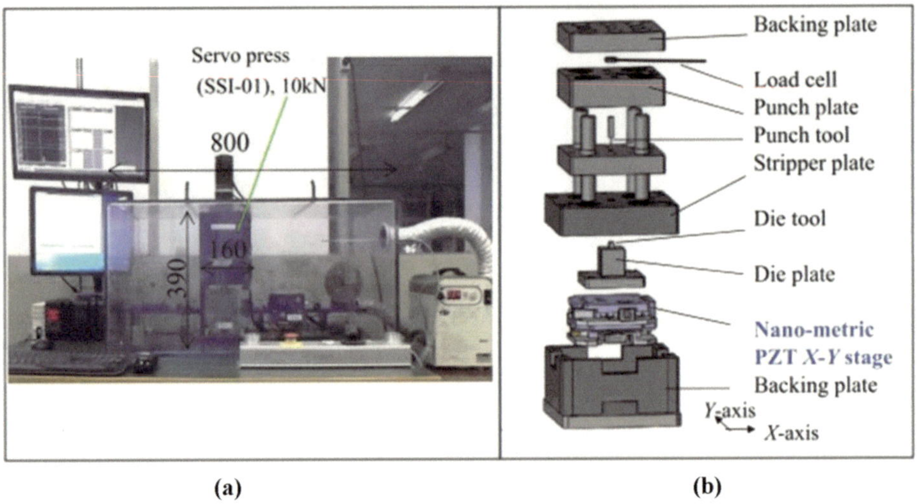

Figure 11.
Aluminum-flamed fine stamping system with high stiffness in the die set and flexible stamping structure. (a) Overview on the CNC mini-stamping with the maximum loading capacity of 10 kN, and (b) illustration of the die set.

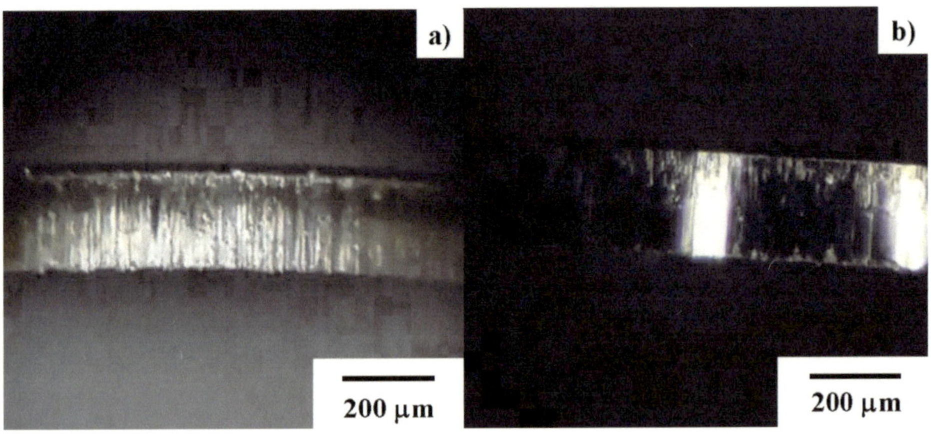

Figure 12.
Comparison of the optical-microscopic image on the pierced AISI316L hole surfaces when using two punches. The edge curvature in both punches is nearly the same as 2 μm. (a) Pierced hole surface by the normal WC (Co) punch with the sharpened edge, and (b) pierced hole surface by the laser-treated diamond-coated WC (Co) punch.

6.2 Product qualification

The WC (Co) punch with the sharpened edge was also used as a reference punch for comparison of pierced hole surface to the present approach. **Figure 12** compares the sheared surfaces by piercing process with the use of WC (Co) punch and trimmed diamond coated one after 100 shots in continuous.

As depicted in **Figure 12a**, AISI316L sheet was punch out with full burnished surface area ratio even when using the WC (Co) normal punch with the sharpened edge. This fully sheared surface had lots of scratches since the side surface roughness of WC (Co) punch was transcribed onto the hole surface during the piercing process. When using the trimmed diamond-coated punch, the pierced hole also has no fractured surfaces in **Figure 12b**. In addition, this surface has a mirror-shining surface condition with tiny scratches only on its top. The length of nanostructured side surface from the punch edge is 200 μm and equal to the sheet thickness.

Irregular texture between the nano-textured and non-textured side surfaces induced these scratch markings. The essential difference in the pierced surface condition comes from the shearing process by the punch with and without the nanotextures on its side surface from its edge.

Let us analyze the pierced hole surface condition by the nanostructured punch from the multi-dimensional view. As shown in **Figure 13**, SEM observation is made from the lowest magnification to the highest one. As seen in **Figure 13a** and **b**, the pierced surface looks smooth without any scratches just in correspondence to the optical-microscopic observation in **Figure 12b**. With increasing the magnification, this smooth surface is found to have nano-stripes as shown in **Figure 13c** to **13e**. This reveals that nanostructures with the period of 300 nm on the diamond coated punch are imprinted onto the pierced hole surface as nano-stripes. **Figure 13f** proves this imprinting of nanostructures to product surface together with piercing the AISI316L sheet.

WLI is also utilized to make nondestructive analysis on the imprinted nanotextures in **Figure 13f**. **Figure 9b** depicts the pierced hole surface profile. In correspondence to nano-stripes in **Figure 13f**, nano-textures are detected on the hole surface. The DEAP alogorism in WLI also biased the measurement of nanotextures on the pierced hole by its local curvature to provide Λ_{hole} = 900 nm in **Figure 9b**. Since $\Lambda_{punch} = \Lambda_{hole}$ = 900 nm by WLI in **Figure 9**, the nanostuctures on the punch side surface is simultaneously imprinted onto the hole surface together with piercing the hole.

In the algorithm of DEAP, the effects of large curvatures on the interferometric measurements are difficult to avoid when analyzing nanotextured periods on the punch and pierced hole surfaces. The nanotextures on the punch and hole surfaces were over-estimated to have larger periods by this curvature effect in the WLI and DEAP analysis.

6.3 Ejection of debris particles

The debris particles splash in the air, easily deposit on the die surfaces and often lock the further steps in cutting and piercing the work materials. In particular, when

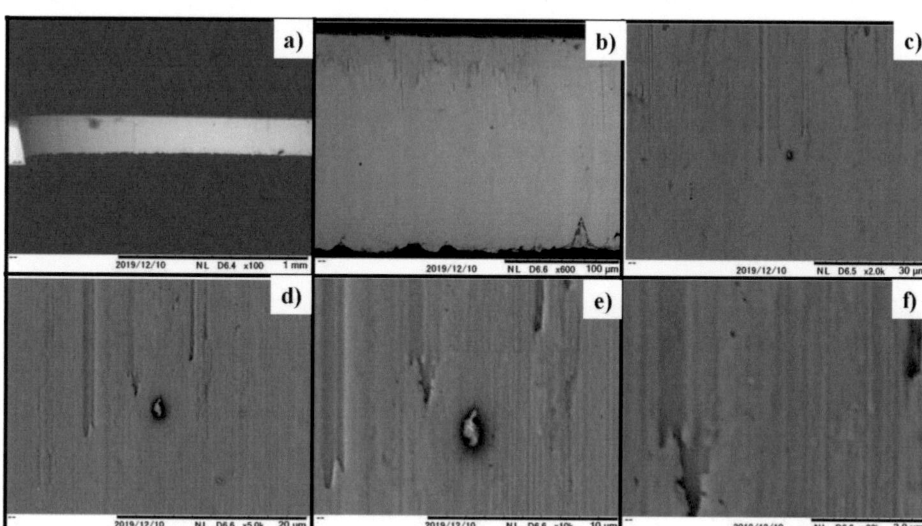

Figure 13.
SEM image on the pierced AISI316L hole surface by using the laser treated diamond-coated WC (Co) punch with varying the magnifications from (a) to (f).

fine piercing the work in the narrow clearance, they deposit on the punch head and side surfaces under high static pressure. This deposition increases the friction and wear in piercing, and damages to the tools. In the lubricated conditions, those particles are trapped into the lubricating oils and ejected to outside of cutting and stamping processes together with liquid lubricants. However, in the case of dry piercing of works, there are no ways to pocket the splashing particles and push out them from the piercing front on the interface between punch and work to its end. Those residual debris particles adhere to the punch surface and lock the piercing process at the risk of severe damage to punch edge and surfaces. Hence, how to trap those debris particles and to eject them out of the piercing system becomes an issue to promote the production quality in fine piercing.

The nanotextures formed on the punch side surface are expected to be working as a nano-groove to trap and eject these debris particles from the vicinity of punch edge to the length of punch. After continuously piercing the AISI316L sheets in a thousand shots, the punch surface was precisely analyzed by SEM. **Figure 14** depicts the punch head and side surfaces with varying the magnification in SEM observation. As seen in **Figure 14a**, no adhesion of debris particles is detected on both surfaces. With increasing the magnification in SEM, the iron-rich debris particles of AISI316L are trapped into the nano-grooves on the side surface as shown in **Figure 14b** and **c**. As depicted in **Figure 14c**, most of nano-grooves trap the debris away from the punch edge by 8 μm.

The nanotextures were formed from the punch edge to the length of 0.2 mm along the punch axis. Consider that this punch is pierced into the AISI316L sheet, and the sheared debris fragments by the sharp punch edge are infiltrated into these nanotextures. As seen in **Figure 14**, less amount of particles is trapped at the vicinity of punch edge but a lot of particles are lodged into them even far from the edge by 100 μm. This suggests that the trapped debris particles are transferred from the punch edge to the punch length during the piercing process. In order to demonstrate this transfer process, the whole punch surface is precisely analyzed along the length of punch. **Figure 15** depicts how the debris particles are trapped and transferred to the punch length (L).

As stated before, little debris fragments are trapped at the vicinity of the punch edge. They fill into the nanostructured grooves; e.g., at L = 8 μm from the edge, most of grooves are stacked by them. To be interested, they overlapped the groove and form an agglomerate of debris fragments at L = 15 μm. Each nano-groove first traps a debris fragment and is gradually packed by debris with increasing the number of shots in piercing. Once the amount of debris exceeds the trapping capacity to fully pack the fragments with the length of 2 μm for each nano-groove, the debris fragments agglomerate to a platelet. These platelets are only seen around L = 15 μm. Since no platelets were seen for L > 15 μm, they delaminate from nanostructured punch surface and transfer to further length of punch.

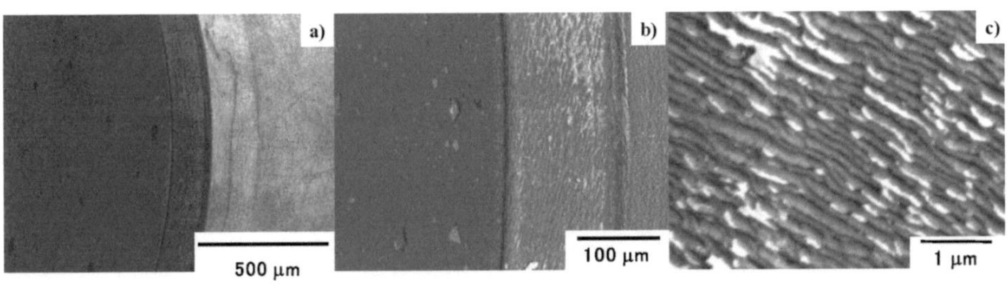

Figure 14.
SEM image on the laser-treated diamond-coated punch near the sharpened edge. (a) Lowest magnification, (b) lower magnification, and (c) higher magnification.

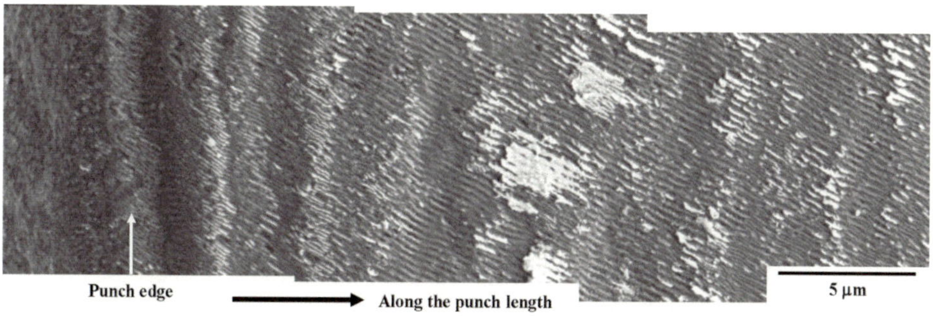

Figure 15.
Fine SEM image on the laser-treated diamond-coated punch from the sharpened punch edge to the end of post-treated zone.

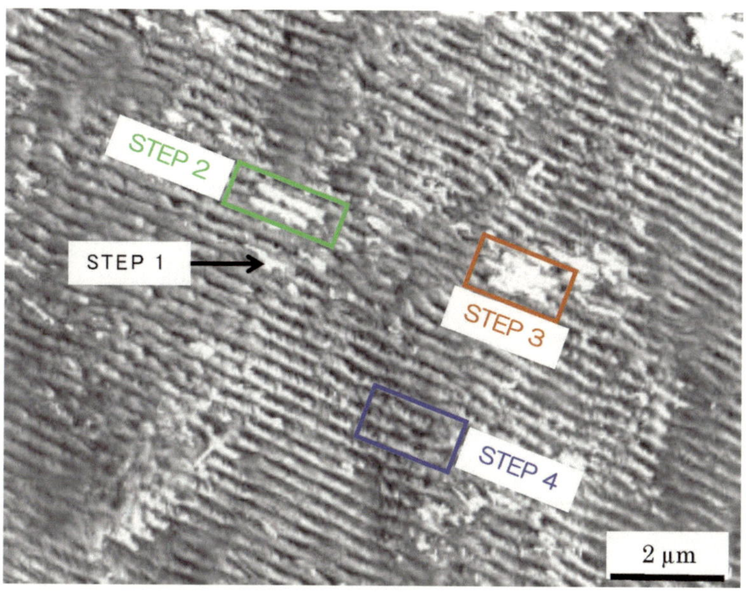

Figure 16.
Ejection model to capture, release and transfer the debris particles from the piercing front to the outside.

Let us consider this ejection mechanism of debris particles in dry piercing. Due to precise SEM analysis on the nano-textured punch surface after continuous piercing in 1000 shots, various steps in this mechanism are described as shown in **Figure 16**.

At the step-1, the AISI316L debris particles are trapped into a single nano-groove. This first trapping of debris occurs on the contact interface of nanostructured punch surface and AISI316L work under high static pressure during every shot in piercing. With increasing the number of shots, the probability also increases for the debris particles to be trapped into two adjacent nano-grooves in the setp-2. In further continuous piercing, each trapped debris agglomerates on a couple of nano-grooves in the step-3. When the size of agglomerates exceeds the critical volume of 5 μm^3, they delaminate by themselves and dislodge to be pushed down into the punch length during the shearing process in piercing. At the step-4, the nano-groove becomes vacant enough to trap new debris particles in further piercing process. This ejection mechanism of debris particles during the piercing is effective to be free from their locking to clearance and to continue the fine piercing process with high product quality.

7. Discussion

CVD-diamond, polycrystalline and single-crystal diamonds have intrinsic hardness ranging from 5000 HV to 10000 HV, controllable electric resistivity by doping from 100 GΩ·m to semi-conductivity, high thermal conductivities around 2200 W/(m·K), and high thermal stability. In its industrial applications to piercing punches and dies, their laser adjustment and surface treatment is indispensable to make full use of these properties.

The original roughness of bare diamond coating by its tetragonal crystal growth is in the order of 3 to 5 μm. As demonstrated in **Figure 3**, its maximum roughness can be lowered to be less than 0.8 μm even by the multi-pulse laser irradiation. A microgroove with the length of 10 mm, the width of 125 μm and the depth of 3.8 μm is accurately cut into the diamond film. This proves that the tailored geometry for fine piercing punch is shaped onto the diamond coating within the tolerance of submicron meter.

When using the femtosecond laser-treatment, the surface roughness is much reduced as shown in **Figure 6** through the laser beam control in **Figure 2b**. This suggests that the punch edge as an intersection of its head and side surfaces can be sharpened by trimming these two surfaces and reducing their surface roughness. **Figure 7** proves that the punch edge sharpening is driven by this surface trimming processes in **Figure 2b**. The most preferable merit to this femtosecond laser treatment is a simultaneous nano-structuring on the punch side surface together with the laser trimming process. As depicted in **Figure 8**, this nanostructuring by LIPSS commences just from the punch edge to the length on its side surface. The period of induced nanostructures is dependent on the laser pulse width, the fluence and the laser beam control in addition to the diamond film surface roughness. Their depth into the diamond is mainly controlled by the fluence in laser trimming, as shown in **Figure 10**. The direction of nanostructures is also tunable by the optical control.

In addition to the skew angled nanostructures in **Figures 8c** and **10**, each nanostructure by LIPSS can be formed in the circumferential and longitudinal directions,

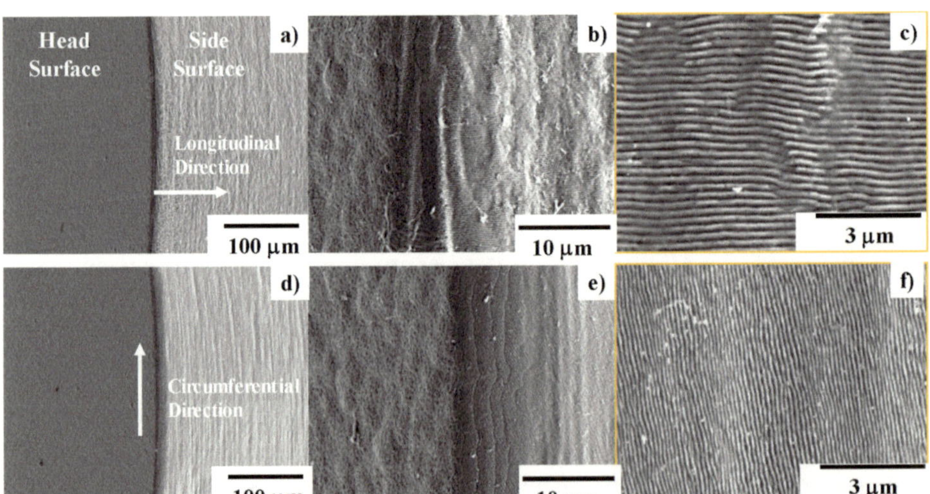

Figure 17.
Controllability of the laser-induced nanostructures onto the side surface of CVD diamond-coated punches. (a-c) A longitudinal alignment of nanostructures on the trimmed punch with varying the magnification in SEM, and (d-f) a circumferential alignment of nanostructures on the trimmed punch with varying the magnification in SEM.

respectively, by the polarization technique [37]. **Figure 17** depicts two typical nanostructure alignments onto the diamond-coated punch side surface. SEM images in **Figure 17a** to **17c** with varying the magnification, depict the longitudinally aligned nano-grooves along the punch length. **Figure 17d** to **17f** show the SEM images on the circumferentially aligned nano-grooves with varying the magnification. The direction of nanostructures as well as their LIPSS periods are modified by this polarization control. In particular, when controlling the formation of nanogrooves in the circumferential direction, the LIPSS-ripples have smaller period of 100 nm than 300 nm for the nanogrooves formed in the longitudinal direction.

This control of nanogroove directions has direct influence on the piercing behavior. As introduced in [38], the sharpened punch edge behaves as a blade to cut into the work at the beginning of the piercing process as illustrated in **Figure 18**. The droop is formed by elastoplastic deformation of work at the indentation of punch, and the sheared work surface is generated by the contact of work to punch side surface before final fracture. In the nanostructured punches, each LIPSS-formed nanostructure works as a blade to advance the shearing process in piercing.

When using the nanostructured punch in **Figure 17a** to **17c**, the straight nano-grooves are imprinted onto the pierced hole surface to have the LIPSS-period of 300 nm. During this piercing process, the debris fragments are easily driven to the length of punch. On the other hand, when using the nanostructured punch in **Figure 17d** to **17f**, the finer nano-grooves with the LIPSS period of 100 nm are formed in the circumferential directions on the pierced hole surface. Every debris particle is stacked in each nano-groove during piercing. These ejection processes with dependence on the nanostructure alignment have an importance role to preserve the high quality piercing of work and prolong the punch life without damages by debris particles.

Let us reconsider the effect of debris particle fragmentation to the piercing process. How to deal with the debris particles is an essential issue in mechanical machining and metal forming. As proposed in [22, 39, 40], the micro-dimples on the rake surface of cutting tools work as a micro-reservoir to stock the debris particles. In case of end-milling of aluminum alloys, the machined chips stack to

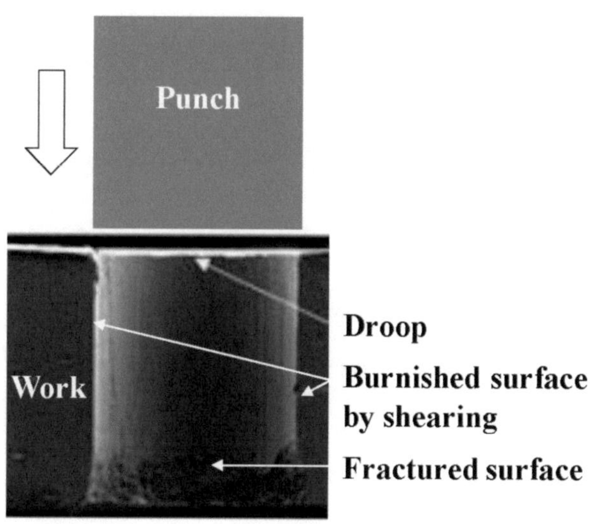

Figure 18.
Schematic view on the cross-section of pierced hole in the AISI316L work by punching. A droop was formed by the initial indentation of punch; then, the burnished and fracture surfaces are formed by this shearing process of ductile work.

these dimples so that the adhesion wear of debris to cutting tools is saved to prolong the cutting tool life. In the metal forming under lubricating oils, the debris particles are included into these lubricating oils and ejected together to outside of forming system. When using this micro-dimple technique, a lubricating oil is indispensable to house and drive the debris particles into the micro-dimples. In case of the dry machining and metal forming, how to trap and eject them out of the working space becomes an issue to prevent the tools and dies from severe damage and to be free from the shortage of their lives. Nano-structuring to cutting and metal forming tools proves a method for trapping and ejecting the debris fragments to outside of the cutting and forming systems together with the movement of tools.

How to preserve the product quality, becomes another issue to be solved by tooling appropriately. Although this problem is not so severe in cutting and machining, a product quality assurance during the metal forming is an essential issue for die and punch design. The product surface quality in piercing depends on the shearing process on the interface between punch surface and work and on the flow stress of work materials. During the piercing process, the initial contact interface starts at the punch edge; the strain concentration at the sharp edge drives the shearing process of work materials. Let us consider how the punch edge profile in the edge width influences on the piercing process.

Two types of punch are prepared for piercing the amorphous electrical steel sheets with the thickness of 25 μm. As shown in **Figure 19**, one is a WC (Co) punch with the sharpened edge width of 2 μm and another is a laser-trimmed diamond-coated punch also with the edge width of 2 μm. The difference in the geometric topology between two punch edges is noticed as an edge profile morphology. The WC (Co) punch has a diffusing edge profile as seen in **Figure 18a** to **18c** while the laser trimmed diamond-coated punch has a homogenous edge profile as shown in **Figure 18d** to **18f**. A brittle amorphous electrical steel sheet is employed to describe the effect of two edge profiles to its piercing behavior.

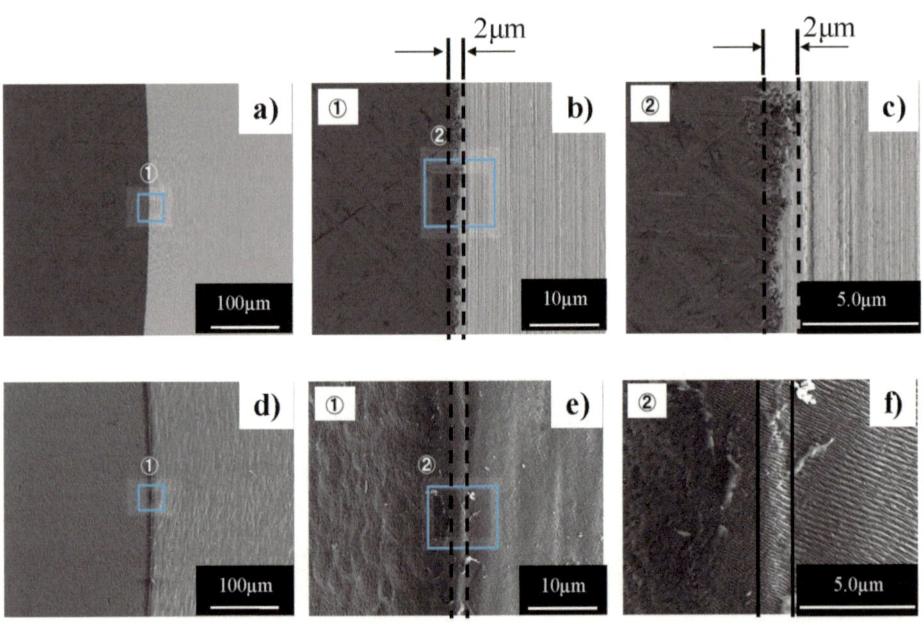

Figure 19.
Comparison of SEM images with different magnifications between two fine piercing punches with the same edge width of 2 mm. (a–c) WC (Co) punch with the sharpened edge width of 2 mm, and (d–f) diamond coated WC (Co) punch with the laser trimmed edge width of 2 mm.

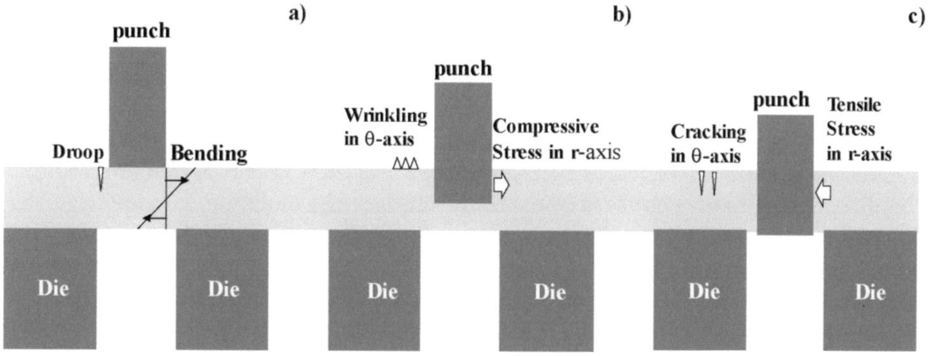

Figure 20.
Damaging process induced into the work sheet by the piercing process from (a) to (c). (a) Formation of a droop with circumferential cracks, generated by the initial indentation of punch to work, (b) formation of circumferential wrinkles, induced by the compressive stress in the radial direction, and (c) formation of circumferential cracks by perforation of a hole.

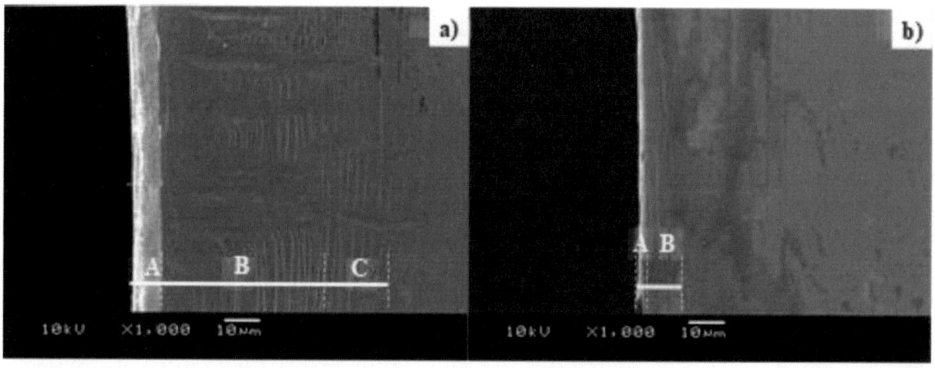

Figure 21.
Micro-damages induced into the amorphous electrical steel sheet by the piercing process. A-zone is a droop with the circumferential cracks at the stage of punch indentation. B-zone is a wrinkle with peaks and valleys where the short surface cracks are seen on the peaks. C-zone includes the long surface cracks in the circumferential direction. (a) When using the WC (Co) punch, and (b) when using the diamond-coated punch.

First in the punching process as depicted in **Figure 20**, a brittle material begins to make elastic shear deformation by initial indentation of a punch edge into it. This bending deformation by indentation of the punch edge, results in the formation of droop with surface cracks in the circumferential direction or in θ-axis. In further indentation of punch, the compressive stress is induced in the radial direction to push back the work material in shearing. Under this compressive stress state, the wrinkling occurs in θ-axis. When punching out, the tensile stress is applied to the work so that the circumferential cracks are generated in the work surface.

Figure 21 compares the pierced hole surface by two punches with the different edge profile. When using the sharpened WC (Co) punch, the droop, the wrinkling, and the circumferential cracking are all seen on the pierced sheet surface as A-zone, B-zone and C-zone, respectively in **Figure 21a**. On the other hand, the droop and the circumferential cracking are measured in **Figure 21b** when using the nano-structured diamond-coated punch. No circumferential wrinkling takes place in the latter. In addition, the circumferential cracking only occurs at the vicinity of hole surface. This difference does not come from the sharp edge width but from the homogeneous edge profile. As seen in **Figure 19**, the edge profile of sharpened WC (Co) punch is diffusing so that the circumferential distortion could be easily induced by the compressive stress at the contact of convex punch edge parts to the work. In the case of the nanostructured punch with homogeneous edge profile, the brittle work sheet is sheared without the wrinkling in the circumferential direction. This difference in the piercing behavior suggests that nano-structured punch has possibility to reduce the induced defects by controlling the structure of nano-grooves such as the direction of nanostructures and their LIPSS-period and depth. The direction of nanogrooves is optimized to reduce the cracking damage; e.g., the longitudinal nanostructuring is recommended to reduce the A- and C-damage widths and to eject the generated debris fragments. The nanostructure depth of 3 to 5 μm is necessary to stimulate the plastic flow of ductile work around the edge profile and to improve the punch life. The LIPSS-period is designed to reduce the A-damage width as well as the piercing stress.

Let us be back to how to imprint the nanostructures by stamping as shown **Figure 11**. In case of the piercing process, the pierced hole surface is macroscopically smooth with metallic shining and microscopically has nanotextures on it. In case of the embossing and coining processes, the tailored nanotextured surfaces are directly imprinted onto various mechanical parts and tools. Nontraditional design on the micro−/nano-textures leads to development of new mechanical elements in application.

In fine piercing operations, most of piercing punches and dies have complex shaped heads and core-cavities with the accurate dimension, respectively. Let us evaluate on the application of the present laser trimming method to fabricate those complex-shaped punches and dies. A cross-lettered WC (Co) punch was employed for femtosecond laser trimming, as shown in **Figure 22a**. The laser-machining path schedule was optimized to make homogeneous machining the whole punch side surfaces around the cross-lettered head. **Figure 22b** shows the SEM image on the vicinity of punch edge with high magnification. The side surface was trimmed and nanostructured to have the LIPSS-period of 300 nm in the similar skewed angle as seen in **Figure 8c**. This demonstrates that the complexed shaped punches and dies are laser-trimmed to sharpen their edges and to form the nano-structures on their surfaces by the present laser-treatment.

In addition to the CVD diamond coatings, this simultaneous laser trimming with nanostructuring is successfully applied to the ceramic coated dies such as CrN, AlCrN and DLC as well as the nitrided and carburized tools. Those nanostructured

Figure 22.
Laser trimming of the complex shaped, diamond-coated WC (Co) punch for fine piercing. (a) Overview of the cross-lettered punch, and (b) nanostructured side surface of punch across its edge. The nanotextures are wavy in nature.

dies and tools work in metal forming to accurately yield the engineered surfaces onto the metallic products with higher cost-performance.

8. Conclusion

A circular economy stands on the sustainable manufacturing with high material efficiency, less emission of wastes and long-life tooling. The CVD diamond coated WC (Co) tooling grows up as a reliable method. The used diamond film was perfectly ashed with less damage to WC (Co) substrate. The tailored WC (Co) substrate is recycled as a tool after recoating the diamond films. Through the laser-treatment of recoated diamond layer, high qualification of products is put into practice together with prolongation of tool life. The laser-treatment provides a reliable method to resize the rough shape of as-coated diamond film into tailored geometry for punch and die in metal forming. No significant damages are generated through this processing. In particular, the femtosecond laser-treatment plays a role to adjust the as-coated diamond punch as a tool for ultra-fine piercing of stainless steel and amorphous electrical steel sheets at first. In second, the nano-textured surfaces are accommodated to punches and dies. In third, the laser-trimmed punch and die lives are prolonged with sufficient cost-competitiveness.

This laser treatment is characterized by the simultaneous edge-sharpening with nanostructuring. Especially, the laser-trimmed punch has a homogeneously sharpened edge with its width less than 2 μm. It has also a nano-structured side surface with the tailored LIPSS-period. In the fine piercing process, this nanostructure plays a double role. The pierced hole surface quality is improved from a fine surface with fully burnished area ratio to the hole surface to an ultrafine surface with mirror-polished condition. The generated debris fragments are ejected from the piercing front to the length of punch through this nanostructured groove. The affected zone width by piercing the ductile work is reduced by tailoring this nanostructures on the piercing punch. Since this zone of work experiences the plastic straining and strain recovery from plastic to elastic states, the reduction of its width improves the quality of products. In case of Fe-Si alloyed electrical steels, this reduction decreases the iron loss in the motor core and increases the product reliability as connector and sensing devices. In the similar manner, the damaged zone width of brittle amorphous sheets is also much reduced by this ultrafine piercing to lower the iron loss.

The imprinting method of nanostructures onto the product surface can be widely utilized to prepare for the engineered surfaces onto various products in application. Nanoscopic angulation onto the micro-textured product surfaces is useful to modify the original hydrophilic metallic surface to super-hydrophobic one. The critical heat flux of heat sink is enhanced by controlling the bubble nucleation at the imprinted micro-/nano-structure surfaces. The regularly aligned micro-/nano-structures surface work as an anti-bacteria part to prevent the human handling from infection. The tailored micro-/nano- textures on the medical tools assist a doctor to pick up and hold the targeting cells and organic parts. Through the imprinting process to dies and tools, various engineered surfaces are tailored and yielded onto the products.

Acknowledgements

The authors would like to express their gratitude to Mr. Y. Kira and Mr. S. Ishiguro (Graduate School of Engineering, University of Toyama) for their help in analysis. This study was financially supported in part by METI-program on the supporting industry, 2020.

Conflict of interest

The authors declare no conflict of interest.

Author details

Tatsuhiko Aizawa[1*], Tadahiko Inonara[2], Tomoaki Yoshino[3], Tomomi Shiratori[4] and Yohei Suzuki[3]

1 Surface Engineering Design Laboratory, SIT, Tokyo, Japan

2 LPS-Works, Co., Ltd., Japan

3 Komatsuseiki Kosakusho Co., Ltd., Japan

4 University of Toyama, Japan

*Address all correspondence to: taizawa@sic.shibaura-it.ac.jp

References

[1] Reineck I., Sjoestrand M. E., Karner J., Pedrazzini M., Diamond coated cutting tools. Int. J. Refractory Metals Hard Metals 14; 1996: 187–193.

[2] Konstanty J., Sintered diamond tools: trends, challenges and prospects. Powder Metallrugy 56(3); 2013: 184–188.

[3] Kuo C., Wang C., Ko S., Wear behavior of CVD diamond-coated tools in the drilling of woven CFRP composites. Wear 398; 2018: 1–12.

[4] Koga N., Usu K., Xu C., Deep drawing of thin stainless steel sheets using diamond tools without lubricant. J. JSTP 53; 2012: 74–78.

[5] Zhao G., Li Z., Hu M., Li K., He N., Jamil M., Fabrication and performance of CVD diamond cutting tool in micro milling of oxygen-free copper. Diamond and Related Materials 100; 2019: 107589.

[6] Aizawa T, Masaki E, Morimoto E, Sugita Y. Recycling of DLC-coated tools for dry machining of aluminium alloys via oxygen plasma ashing. Mechanical Engineering Research. 2014; 4(1): 52–62.

[7] Aizawa T, Sugita Y. High-density plasma technology for etching and ashing of carbon materials. Research Reports SIT. 2011; 55(2): 13–22.

[8] Aizawa T, Masaki E, Sugita Y. Oxygen plasma ashing of used DLC coating for reuse of milling and cutting tools. In: Proc. Int. Conf. Mater; Advanced Process Technology; 2011: 15–20.

[9] Aizawa T, Fukuda T. Micro-texturing onto carbon-based coatings via oxygen plasma etching. Research Reports SIT. 2012; 56(2): 66–73.

[10] Aizawa T, Fukuda T. Oxygen plasma etching of diamond-like carbon coated mold-die for micro-texturing. Surf. Coat. Technol. 2013; 215: 364–368.

[11] Aizawa T, Masaki E, Sugita Y. Complete ashing of used DLC coating for reuse of the end-milling tools. Manufacturing Letters. 2014; 2: 1–3.

[12] Aizawa T., Micro-manufacturing by controlled technologies. J. Automotive Engineering. 2018; 72 (6): 35–41.

[13] Aizawa T., Controlled post-treatment of thick CVD-diamond coatings by high-density plasma oxidation. In Ch. 4; Chemical Vapor Deposition for Nanotechnology, Intech Open 2019: 75–94.

[14] Yamauchi K., Aizawa A. High density plasma ashing of used diamond coated short-shank tools without damage to WC (Co) teeth. Proc. 11th ICOMM 2016; 50: 1–5.

[15] Yunata E. E., Aizawa T., Yamauchi K. High density oxygen plasma ashing of CVD-diamond coating with minimum damage to WC (Co) tool substrates. Mechanical Engineering. JSME 2016; CD-ROM.

[16] Yunata E. E., Aizawa T., Micro-grooving into thick CVD diamond films via hollow cathode. Manufacturing Letters. 2016; 4: 17–22.

[17] Roy S., Das M., Malik A.K., Balla V. K., Laser melting of titanium – diamond composites: Microstructure and mechanical behavior study. Mater. Lett. 2016; 178: 284–287.

[18] Popovich A. F., Ralchenko V. G., Balla V. K., Malik A. K., Khomich A. A., Growth of 4″ diameter polycrystalline diamond wafers with high thermal conductivity by 915 MHz microwave

plasma chemical vapor deposition. Plasma Sci. Technol. 2017; 19 (3): 035503.

[19] Dowson D., Dalmaz G., Childs T.H. C., Taylor C. M., Godet M., Wear Particles: From the Cradle to the grave. 1st Ed. 1992 Elsevier.

[20] Kirk A. M., Shipway P. H., Sun W., Bennett C. J., Debris development in fretting contacts - debris particles and debris beds. Tribology International. 2020; 149: 105592.

[21] Aizawa T., Satoh T., Shiratori T., Micro-joining of shaped stainless steel sheets for fuel injection orifice with high misting capability. Proc. 22nd Int. ESAFORM Conf. Mater. Form. AIP Conf. Proc. 2019; 2113: 050007–1.

[22] Aizawa T., Inohara T., Wasa K., Femtosecond laser micro−/nano-texturing of stainless steels for surface property control. J. Micromachines, MDPI 2019; 10–512: 2019: 1–12.

[23] Aizawa T., Wasa K., Tamagaki H., A DLC-punch array to fabricate the micro-textured aluminum sheet for boiling heat transfer control. J. Micromachines, MDPI 2018; 9–147: 1–10

[24] Aizawa T., Ono N., Boiling heat transfer control by micro−/nano-texturing of metallic heat-spreading devices. Proc. WCMNM2021 (in press).

[25] Shiratori T., Yoshino T., Suzuki Y., Nakano S., Yang M., Effects of ion processing of cutting edge of tools on tool wear and punching characteristics in mciropunching process. J. JSTP 2019; 60–698: 58–63.

[26] Enomoto Y., Soma K., Development of the amorphous motor balancing both resource saving and energy saving. J. Soc. Mech. Eng. 2014; 117: 753–756.

[27] Aizawa T., Inohara T., Multi-dimensional micro-patterning onto ceramics by picosecond laser machining. Res. Rep. Shibaura Institute of Technology 2012; 56 (1): 17–26.

[28] Aizawa T., Inohara T., Geometric adjustment and sizing of CVD-diamond coatings via oxygen plasma etching and laser machining. Proc. 7th SEATUC Conference (Bandon, Indonesia) 2013: 141–146.

[29] Aizawa T., Inohara T., Pico- and femtosecond laser micromachining for surface texturing. In Chapter 3 of Micromachining, InTechOpen, London, UK 2019: 35–62.

[30] Aizawa T., Shiratori T., Inohara T., Short-pulse laser precise trimming of CVD-diamond coated punch for fine piercing. Proc. 2nd Asian Pacific Symposium on Technology of Plasticity (APSTP) 2019: 123–128.

[31] Aizawa T., Shiratori T., Yoshino T., Inohara T., Femtosecond laser trimming of CVD-diamond coated punch for fine embossing. Mater. Trans. 2020; 61 (2): 244–250.

[32] Aizawa T., Shiratori T., Kira Yoshihiro, Inohara T., Simultaneous nano-texturing onto a CVD-diamond coated piercing punch with femtosecond laser trimming. Appl. Sci., MDPI 2020; 10, 2674: 1–12.

[33] van Driel H.M., Sipe J.E., Young J.F., Laser-induced periodic surface structure on solids: A universal phenomenon. Phys. Rev. Lett. 1982;49: 1955–1958.

[34] Aizawa T., Inohara T., Wasa K., Fabrication of super-hydrophobic stainless steel nozzles by femtosecond laser micro−/nano-texturing. Int. J. Automation Technol. 2020; 14 (2): 159–166.

[35] Prawer S., Nemanich R. J., Raman spectroscopy of diamond and doped diamond. Phil. Trans. Royal Soc. A

Mathematical Physical and Engineering Sciences. 2004; 362(1824): 2537–2565.

[36] Larkin K. G., Efficient nonlinear algorithm for envelope detection in white light interferometry. J. Opt. SoC. Am. 1996; 13 (4): 832–843.

[37] Tsubakimoto K., Miyaji G., Sueda K., Miyanaga N., Application of wavefront and polarization technique for power laser. Jpn. J. Optics. 2006; 35 (12): 635–642.

[38] Kira Y., Aizawa T., Suzuki Y, Shiratori T., Effect of punch edge profile to the piercing behavior of electrical amorphous steel sheet stacks. Proc. Autumnal Meting of JSTP 2020: 171–172.

[39] Sugihara, T., Enomoto, T., Performance of cutting tools with dimple textured surfaces: comparative study of different texture patterns. J. Int. Soc. Precision Engineering Nanotechnology. 2017: 49; 52–60.

[40] Aizawa, T., Morita, H., Tribological design and engineering in surface treatment for semi-dry and dry stamping. Proc. ICTMP2016. 2016: 14–28.

Chapter 5

Diamond-Like Carbon (DLC) Coatings for Automobile Applications

Funsho Olaitan Kolawole, Shola Kolade Kolawole, Luis Bernardo Varela, Adebayo Felix Owa, Marco Antonio Ramirez and André Paulo Tschiptschin

Abstract

Diamond-like carbon (DLC) coatings are amorphous carbon material which exhibits typical properties of diamond such as hardness and low coefficient of friction, characterized based on the sp^3 bonded carbon and structure. The proportion of sp^2 (graphetically) and sp^3 (diamond-like) determines the properties of the DLC. This coating can be applied to automobile engine component in an attempt to provide energy efficiency by reducing friction and wear. However, DLC coatings are faced with issues of thermal instability caused by increasing temperature in the combustion engine of a vehicle. Therefore, it became necessary to seek ways of improving this coating to meetup with all tribological requirements that will be able to resist transformational change of the coating as the temperature increases. This chapter discusses the need for diamond-like carbon coatings for automobile engine applications, due to their ultra-low friction coefficient (<0.1) and excellent wear resistance (wear rate ~ 7×10^{-17} m^3/N.m). The importance of DLC coatings deposited using PECVD technique, their mechanical and tribological properties at conditions similar to automobile engines would also be discussed. Non-metallic (hydrogen, boron, nitrogen, phosphorus, fluorine and sulfur) or metals (copper, nickel, tungsten, titanium, molybdenum, silicon, chromium and niobium) has been used to improve the thermal stability of DLC coatings. Recently, incorporation of Ag nanoparticles, TiO_2 nanoparticles, WO_3 nanoparticles and MoO_3 nanoparticles into DLC has been used. The novel fabrication of diamond-like carbon coatings incorporated nanoparticles (WO_3/MoO_3) using PECVD for automobile applications has shown an improvement in the adhesion properties of the DLC coatings. DLC coatings had a critical load of 25 N, while after incorporating with WO_3/MoO_3 nanoparticles had critical load at 32 N and 39 N respectively.

Keywords: automobile, diamond-like carbon, nanoparticles, PECVD, tribological

1. Introduction

In an attempt by the automobile industry to cut down on the consumption of fuel in the combustion engine of automobile vehicle, it is important to note that reduction in friction and wear plays an important role in reducing the energy consumed and ensuring energy efficiency. It has been reported by Holmberg et al. [1] that approximately 11.5% energy is required to overcome friction in a combustion engine of a passenger car as seen in **Figure 1** [1]. To reduce the energy required to overcome friction in a combustion engine of a passenger car it is necessary to provide coatings that possess very low coefficient of friction and has high resistance against wear. DLC coatings have been known for their low coefficient of friction and excellence tribological properties in automobile engine parts application [1]. The hardness of DLC coatings range from 5 to 40 GPa, low friction coefficient value (<0.1) and high wear resistance (~3 x 10^{-14} m^3/N.m) [2]. Although these properties may vary, depending on deposition technique such as Physical Vapor Deposition (PVD) and Chemical Vapor Deposition (CVD) which are the two (2) main classes of deposition techniques used [3]. PECVD provides durable DLC coatings, making them important for coating automobile engine parts to withstand harsh operational conditions [4].

The major problem that may possibly arise during the application of DLC coatings on automobile engine parts at operating conditions is thermal instability. The operating temperature in usually 80 to 200°C, at this temperature the hardness reduces due to transformation of sp^3 carbon to sp^2 carbon, leading to a softer DLC which may easily delaminate and wear out [4, 5]. To improve the thermal stability doping (non-metal/metals) of DLC coatings has been used. Non-metals such as hydrogen, nitrogen, silicon, fluorine and sulfur, and metals such as tungsten, titanium, aluminum, chromium and molybdenum has been used to dope DLC films in an attempt to improve the DLC film [4, 5]. This has improved mechanical and tribological properties of DLC coatings by increasing the sp3 carbon bond. However, the internal residual stress in the DLC coatings becomes very high above 1 GPa, causing poor adhesion to metallic substrate. The internal residual stress is as a result of interfacial and structural mismatch between the DLC coating and doped element. In attempt to solve this problem annealing has been used to reduce the residual stress in the DLC and improve the thermal stability of DLC coatings up to 500°C [6].

Recently, the incorporation of nanoparticles into DLC has been used by first dispersing the nanoparticles into hexane or isopropanol, thereafter, introducing the dispersed nanoparticles into the reactor chamber through the nanoparticle inlet, this deposition method has been described by some authors in literature [7–13]. These nanoparticles which have been used includes Ag [7], TiO_2 [8–10], WO_3 [11, 12] and MoO_3 [13]. In this chapter the use of diamond-like carbon coatings for automobile engine applications, due to their ultra-low friction coefficient and

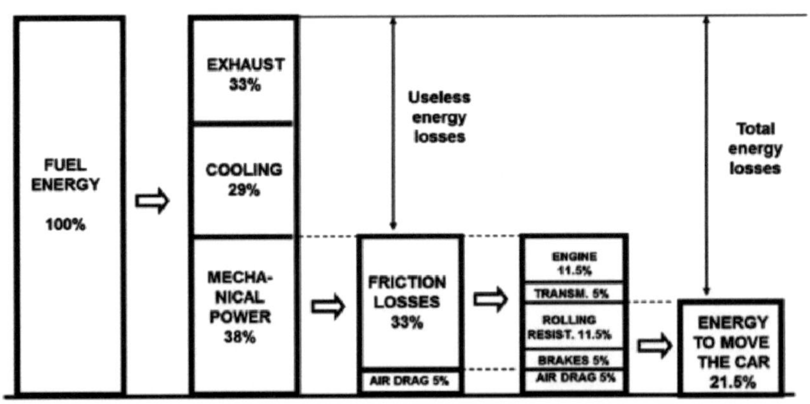

Figure 1.
Breakdown of passenger car energy consumption [1].

excellent wear resistance will be discussed. PECVD deposition technique for DLC coatings, mechanical and tribological properties at conditions similar to automobile engines would also be discussed. Discussion of the thermal stability of DLC coatings and the improvement of the DLC by non-metallic/metallic doping would be made. Lastly, the need to develop novel DLC coatings that can improve thermal stability by incorporating nanoparticles and future DLC coatings.

2. Diamond-like carbon coatings

Diamond-like carbon coatings are amorphous carbon material that exhibit some of the typical properties of diamond and are characterized based on the content of sp^3 bonded carbon and structure [14, 15]. DLCs are usually deposited using precursors of carbon such as ethane (C_2H_6), methane (CH_4), acetylene (C_2H_2), iso-butane (C_4H_{10}), propane (C_3H_8), benzene (C_6H_6), and n-butane (C_4H_{10}) [16]. DLC coatings has the potential to be used on automobile part's interface due to ultra-low friction. Yasuda et al. [17] has used DLC coatings on automobile part and their work showed reduction in friction for PECVD DLC coatings. According to Louda [18] the use of thin coatings in automotive parts brings about eco-friendly environments and economic savings.

2.1 Types of diamond-like carbon coatings

Diamond-like carbon (DLC) coatings varies in mechanical, physical and tribological properties depending on their type. DLC coatings are used in automotive engines due to their ultra-low friction, increase durability, and chemically inert nature. The proportion of sp^2 (graphetically) and sp^3 (diamond-like) determines the properties they possess. Although, such properties can usually be influenced by adding non-metals (hydrogen, nitrogen, silicon, fluorine and sulfur) or metals (tungsten, titanium, aluminum, chromium and molybdenum) as shown in **Figure 2** [4, 5].

2.2 Deposition of DLC film with plasma enhanced chemical vapor deposition

Plasma enhanced chemical vapor deposition (PECVD) is the recently used technique to develop DLCs that are able to reach into the interior of the phase diagram [19] as represented in **Figure 3**. The ternary diagram majorly consists of differing ratios of sp^2 and sp^3 bonded carbon with differing levels of hydrogen. The ratio of these three components can be varied to provide a range of different properties.

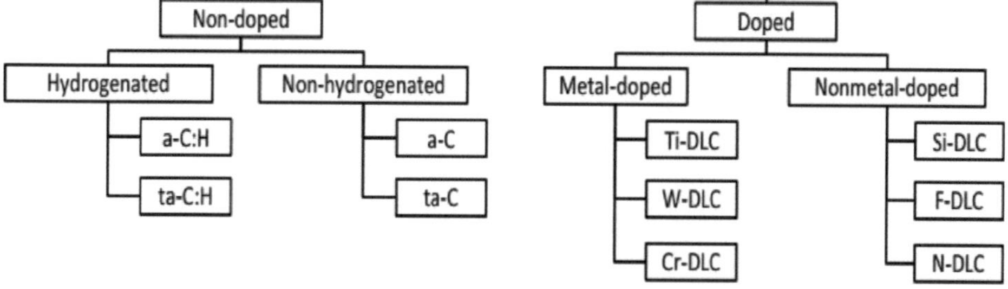

Figure 2.
Classification of DLC coatings [5].

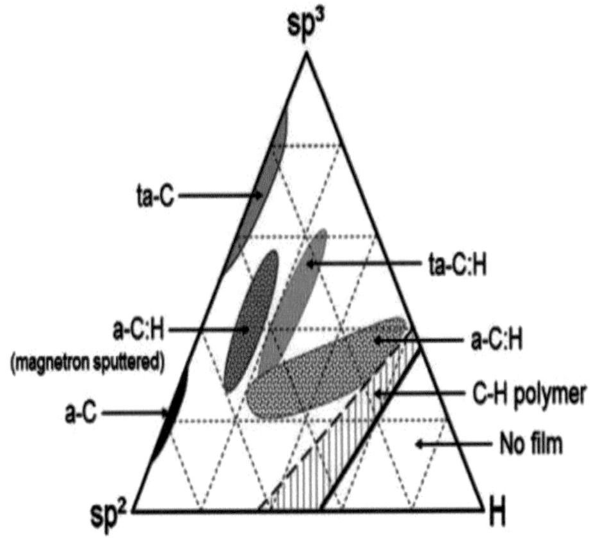

Figure 3.
Ternary phase diagram for various forms of diamond-like carbon [5, 21].

Sp^2 bonding is typical graphite, having a carbon molecule with one double bond to a second carbon atom or to another atom and two single bonds to another atom, leading to planar configuration. While, sp^3 bonded carbon atoms have three single bonds leading to a tetrahedral configuration. Coating names are often abbreviated to ta-C (tetrahedral amorphous carbon) for coatings with very high (almost exclusively) sp^3 content, a-C (amorphous carbon) for coatings with high sp^3 content, between 40 and 80%, and a-C:H (amorphous carbon—hydrogenated) for samples with a greater amount of hydrogen, often coupled with an increase in sp^2 [19]. PECVD technique is based on glow discharge processes supplying hydrogen gases such as acetylene (C_2H_2), with a negatively biased substrate working at radio frequencies (13.56 MHZ) or mid frequencies (10 to 100 KHZ). It is important to note that DLC coatings can be deposited at low substrate temperatures (<200°C). High power of ion acceleration (100 KHZ - 13.56 MHZ), the low cost of equipment and growth of films are the advantages when using PECVD technique [20]. Radio frequency is the most common source of discharge for PECVD deposition [21]. However, some researchers found that using the PECVD technique with DC source pulsed, provides better results in terms of adherence, low friction coefficient, relatively reduced internal stress, high hardness and lower production cost when compared to RF-PECVD techniques [22].

PECVD deposition technique is based on the growth of DLC films using cold plasma, which is characterized by a lack of thermodynamic equilibrium, the ion temperature is lower than the electron temperature. Thus, the kinetic energy (as a result of the temperature) of electrons is much higher than that of ions. The plasma, has a degree ionization in which the gas consists mainly of neutral species excited. The deposition chamber in the PECVD reactor has two electrodes of different areas. The cathode (smallest area) is connected to a current source (Pulsed DC or radio frequency), and the potential difference is applied to it, or polarization voltage. The anode (biggest area) is made up of the walls of the chamber, or by a plate parallel to the cathode, and is grounded. A representation schematic can be seen in **Figure 4**. Before deposition begins, evacuation from the chamber is carried out, a gas is introduced and becomes the medium in the discharge which is initiated and sustained through the current source [22]. The electrons acquire and lose energy

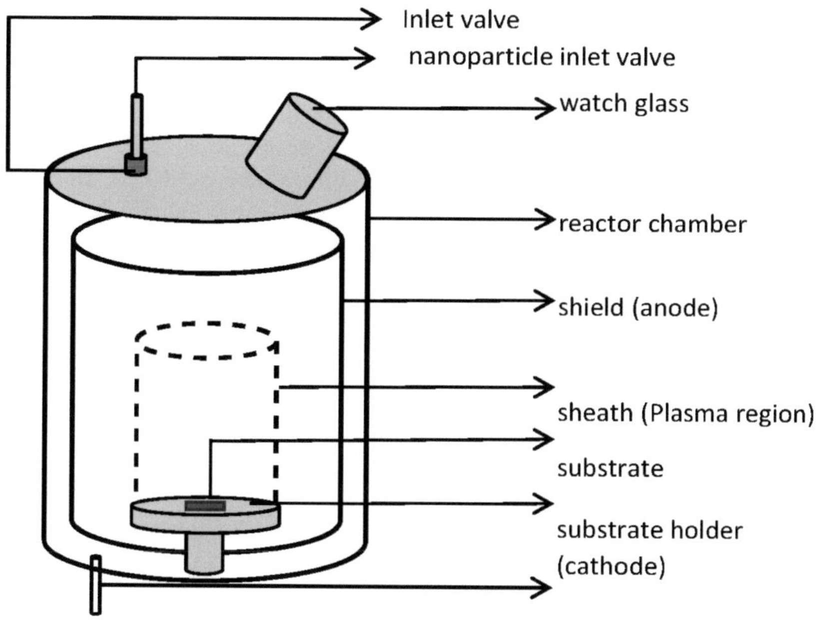

Figure 4.
Schematic for pulsed-DC PECVD with nanoparticle inoculation chamber.

quickly in a sequence of collisions, until they acquire enough energy to ionize (decouple) the gas molecules and produce secondary electrons by impact reactions. As plasma stabilization takes place, electrons are lost in the walls and in electrodes, and the flow is maintained through reactions with secondary electrons and the impact of positive ions on the electrodes. During the stable phase of the plasma, the number of electrons generated and lost is in balance and its stability is directly related to pressure, which influences the free path medium. Plasma being generated by a process of collisions between electrons free accelerated by the electric field and the atoms/molecules of the atmosphere precursor, the basic deposition mechanism involves the creation of species reactive substances such as electrons, ions, molecules, neutral and ionized radicals, causing new ionizations. These reactive species, activated by the discharge, tend to interact with the surface, chemically adsorbing and forming film [22].

The PECVD technique from a pulsed DC source consists of a discharge in low pressure plasma using a pulsed switched source for the plasma generation [23, 24]. Through this technique, different films type (a-C: H and ta-C: H) can be obtained. The structure of the films obtained is composed of the sp^2 hybridization clusters inter-connected by carbons with sp^3 hybridization. Several mechanisms are involved in the deposition of films a-C: H, and the strong dependence on the properties of a-C: H films deposited by PECVD with the polarization voltage (bias-voltage) and the bombardment ions, indicate that the latter have a fundamental role in the deposition of films [25], which makes it necessary to description of the physical process of ionic sub-implantation.

2.3 Application of DLC coatings on automobile engine components

Using DLC as coatings for automobile engine parts can assist to reduce friction and wear, which cannot be achieved by the use of lubricants alone, ensuring an improvement in engine performance and transmission components [26]. Lawes et al. [4] reported diamond-like carbon coatings to have assisted in reduction of fuel consumption (total annual global fuel consumption reduced from 631,109 to 400,109 l/a.) in automobile engines by reducing friction loss due to its self-lubricating properties, chemical stability and weight reduction. The excellent tribological properties (low friction, wear resistance, corrosion resistance, high hardness and chemical stability) of DLC, since their discovery in 1971 has made them widely used in the automotive industry. Different types of DLC films vary in mechanical, physical and tribological properties which is usually determined by their sp3 and sp2 contents in the DLC coating [27, 28]. The mechanical and tribological properties have been analyzed by nano-indentation, scratch, wear and friction testing of an instrumented cam-tappet testing rig to study the tribological properties of DLC coatings for engine applications [15, 16, 29]. Success has been recorded over recent years with the application of DLC coatings to a number of automotive engine components (piston, tappet, camshaft, piston rings, gudgeon pin, valve stem and head, and rocker arm) [4, 16]. DLC coatings applied to automobile parts possess thickness ranging between 1 and 4 μm [4]. DLC coatings in engines needs careful selection of required surface finishing, mechanical properties, and tribological behavior of the coatings with engine oils [4]. The coatings hardness and stiffness are needed to determine the wear resistance of coatings, while friction is controlled by hardness, topography of the surface and tribo-chemical interactions of the dopants with the DLC coatings [4].

2.4 Tribological and mechanical properties of DLC films

High hardness, wear resistance and low friction coefficient are significant for tribological-mechanical properties of the DLC, for a wide range of applications in tribo-systems [30]. Different types of industries (aerospace, automobile, biomedical, mechanical and electronics) have used DLC films, with the intension of increasing the useful life of components and performance of mechanical systems. It can be used as a solid lubricant in environments in which liquid lubrications are not needed, such as in the space environment, food industries and clean environments, conditions of contact with high mechanical loading. Thus, the investigation of the correlation between specific tribological properties of the DLC and working conditions, such as contact pressure, sliding speed, rolling condition, lubrication condition, are very important, and can provide useful information that can aid in predicting tribological behavior of DLC coatings applied to certain machine elements [30].

2.4.1 Adhesion

Adherence is work necessary to separate atoms or molecules at the interface [30]. DLC films are usually faces difficulty with adhesion, due to the high internal compressive stress, accompanies with the growth of the films. This directly interferes with the adhesion between film and metallic substrate, causing detachment of the film [31]. The total tension of the films after the formation and deposition corresponds to a sum of stress effects thermal and intrinsic tension. The thermal stress arises from the difference between the coefficient of thermal expansion of both film and substrate materials, while the intrinsic stress is attributed to the cumulative effect of failures that appear internally in the film during the formation processes [32]. Several methods have been used to decrease internal tension and increase the adhesion of the DLC film on metallic substrates. One of these methods consists of insertion of doping elements such as Ti, Cr, W, N, B and Si, to avoid diffusion of carbon into the metallic substrate, the doping elements form carbides, reducing the total tension in the doped films of DLC [33]. The deposition of DLC films with silicon interlayers using the PECVD Pulsed DC technique has been found to improve adhesion with low coefficient of friction, and lower internal stress which is as a result of interfacial and structural mismatch between the coating and the substrate [21]. Several techniques are used in an attempt to measure film adhesion such as scratch tests or sclerometry (scratch test), scraping test, bending test, impact, cavitation and Rockwell impression [34]. **Figure 5** displays a schematic representation of the scratch test, and the tracks obtained in a test with constant load and with progressive load. The test of sclerometry is a semi-quantitative method that consists of streaking the sample using an indenter, usually diamond. This test allows the determination of properties such as hardness, roughness and specific energy [35].

2.4.2 Wear resistance

Wear is the progressive loss of material from a solid body due to mechanical action (contact and movement of a solid body against a solid, liquid or gaseous body) [36, 37]. In analyzing the wear resistance of DLC films, the hardness of the film or the surface is the first correlation to be made (wear coefficient is inversely proportional to hardness) [36, 37]. The hardness of the film is dependent on the structure and will be higher, with higher concentration of the sp^3 bonds. The wear on DLC coatings is due to two mechanisms namely: friction wear, (related to surface roughness); and wear through transfer of layers through the formation of a tribo-film (transfer layer) induced by contact pressure, which is responsible for decreasing the friction coefficient [20]. The hardness and roughness of the substrate are also factors that affect the wear of the DLC. On low hardness substrates, high plastic deformations occur, so that the film becomes fragile, causing it to break. Schematic of wear configuration (a) pin on disk (b) ball on disk is shown in **Figure 6**.

2.5 Effect of annealing temperature on the thermal stability of DLC coatings

DLC coatings are commonly known for their high internal stresses, which affects their tribological properties and thermal stability. Annealing is mostly used to reduce the internal stress in DLC coatings. Annealing up to 800°C for ta-C is possible and reduces internal stress to an insignificant value (**Figure 7**), while annealing a-C above 500°C reduces the thermal stability. During annealing a cluster of sp^2 bonded carbon atoms will occur, which does not affect the sp^3 content [38, 39]. This may change some of the physical properties, such as optical gap, electrical resistivity, and reduce the compressive stress [38]. At higher temperatures graphitization (sp^3 to

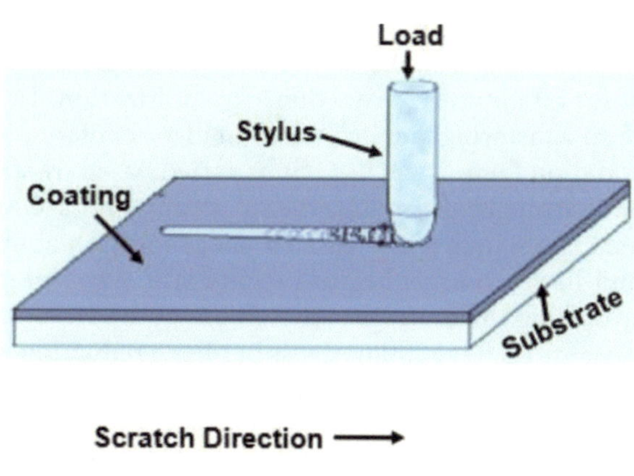

Figure 5.
Schematic representation of the scratch test.

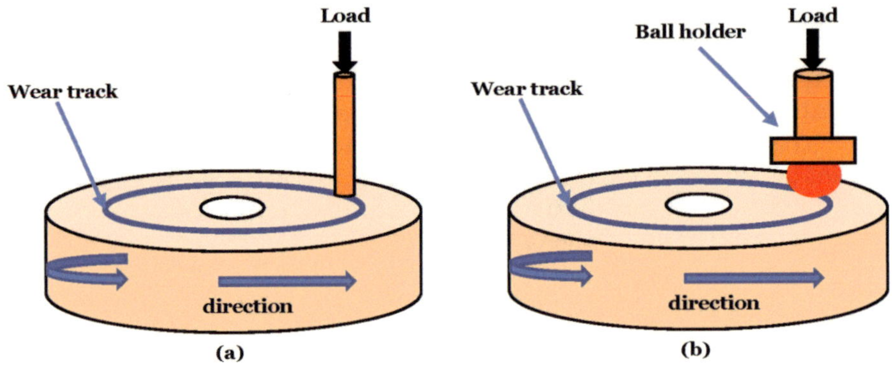

Figure 6.
Schematic of wear configuration (a) pin on disk (b) ball on disk.

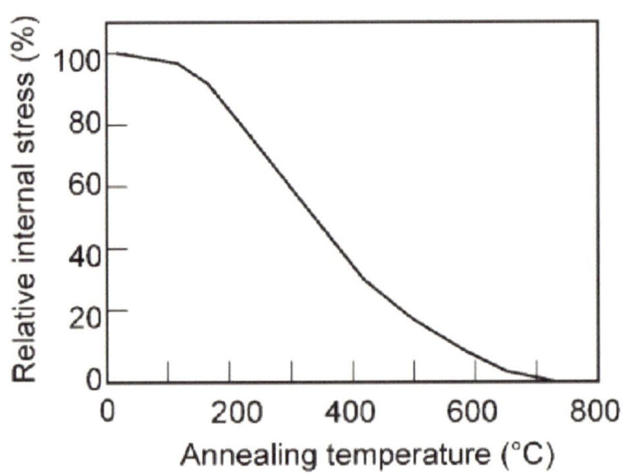

Figure 7.
Relative internal stress against annealing temperature for ta-C coatings [6].

sp^2 transformation) begins, leading to a softer coating which will easily delaminate. Different amorphous DLC films deposited by different deposition methods possess different sp^3 fraction at different graphitization temperature [40].

It has been reported that ta-C hardness does not reduce until about 600°C under vacuum condition [41]. The thin film thickness (nm) remain stable under vacuum condition till 400°C having little graphitization [42]. An investigation of thin films (1.5 nm) revealed changes in the Raman peak position and ID/IG of samples annealed at 300°C for 60 minutes are not due to graphitization. Ta-C film annealed at 300°C has more sp^2 clustering and no reduction in sp^3 content [40]. The starting temperature of oxidation for a-C and ta-C films varied based on structure and the oxidation behavior is strongly affected by the sp^3 content, for a-C and ta-C films [43]. DLC has been known in the past for their poor thermal stability above 500°C, where sp^3 (diamond-like) structure begins to transform to sp^2 (graphitic-like) structure, thereby making the coatings softer. At about the same temperature the loss of coating thickness is above 100 nm, this implies a reduction in coating thickness as the temperature increases. Normally, breakdown and structural collapse of DLC coatings occurs at high temperature above 500°C, which may be attributed to breaking of C–H bonds. Leading to the C–C networks becoming more graphite-like to facilitate the formation of volatile C–O and metal oxide phases. Micro-wear of DLC film occurs as the annealing temperature increases above 200°C, as a result of mechanical stress and thermal degradation of DLC as reported by Lee et al. [44]. Whenever there is a transformation to sp^2 the mechanical strength degrades significantly, which consequently leads to critical failure or engine parts malfunction.

2.6 Modified DLC films by incorporation of additional elements

Modified or doped DLC films according to VDI 2840 standard [45] are denoted as a-C:H:X where X represents non-metallic elements, while, a-C:H:Me where Me represents metallic elements. Until recently, the incorporation of various nanoparticles into DLC has been also used to improve the adhesion properties of DLC by producing nanoclusters that are carbides. In the mid-1970s and 1980s various researchers reported silicon containing a-C:H:Si films deposited using glow discharge decomposition of silane (SiH_4) and hydrocarbon gases (acetylene and methane) and also using d.c. magnetron sputtering [5]. Generally, both a-C:H:X and a-C:H:Me improve mechanical and tribological properties of coatings due to reduction in the residual stress. Increasing the non-metallic or metallic contents in a-C:H, increases the temperature at which structural transformation will occur for films deposited using r.f. PECVD by increasing the sp^3 bonded carbon stabilizing the carbon network, leading to a reduction in graphitization. Hardness and friction coefficients remained nearly constant even after annealing at 500°C in air, whereas, a-C:H coatings failed at 400°C [5].

2.6.1 Metal and non-metal doped DLC coatings for automobile application

Metal-doped DLC (Me-DLC) coatings usually exhibit higher thermal stability than non-doped DLC up to 500°C, which was revealed by X-ray diffraction, transmission electron microscopy, and Raman spectroscopy. Annealing temperature above 500°C, losing high amount of hydrogen from the Me-DLC coatings, causing breakdown and structural collapse of the coatings at high temperature [46]. Non-doped DLC coating has low resistance to wear in lubricating oils containing Molybdenum Dithiocarbamates (MoDTC), this is because of the decomposition and chemical reactivity leading to the formation of oxides and nano-crystallites [47]. Si-doped DLC coating produces anti-wear film, which is usually stronger when the lubricant contains additives. Raman spectroscopy was used to analyze the tribochemical activity of the DLC coatings lubricants with additives. DLC coatings has been used for gear teeth with a coating thickness (1 μm) and deposition temperature (200°C), needing no additional surface finishing [48, 49]. The appropriate method to determine thermal stability of DLC at high temperature is the use of tribological testing equipment such as an Optimol SRV® v4 device which will determine the wear rates and coefficient of friction at high temperature range. Raman spectroscopy, X-ray Photoelectron Spectroscopy (XPS), and nano-scratch experiments can further be used to investigate the mechanism that affect the thermal stability of DLC during their use in internal combustion engines. It is important to note that the Optimol SRV® v4 device can operate between 25 and 900°C.

DLC doped with non-metals (hydrogen, boron, nitrogen, phosphorus, fluorine and sulfur) or metals (copper, nickel, tungsten, titanium, molybdenum, silicon, chromium and niobium) [50] has brought improvement in the thermal stability of DLC coated engine parts. The use of silicon to doped DLC comprising amorphous nano-composites can enhance the thermal stability of DLC. However, a reduction in film thickness occurs at high annealing temperature. Metal-doped DLC creates a two-dimensional array of metallic nano-crystalline clusters and metallic carbides precipitates embedded in an amorphous carbon matrix and reduces surface tension. The use of metal doped DLC helps in stabilizing DLC at high temperatures by delaying graphitization when metallic carbides precipitates are formed [51].

Tungsten and Molybdenum are most commonly used DLC dopant for coating automobile engine parts for improving thermal stability.

2.6.1.1 Tungsten and molybdenum doped DLC coatings

Tungsten and molybdenum have been reported to have been used to doped DLC films in order to reduce friction and wear rate during high temperature [52, 53]. It is important to note the both tungsten and molybdenum will form carbide precipitates when doped with diamond-like carbon coatings [51, 53]. A study of the tribological behavior of DLC-W on Aluminum alloy has been carried out under lubricated condition showed low COF (0.11) and wear rate (0.51×10^{-5} mm^3/Nm), which was due to the presence of tungsten disulphide (WS_2) [54]. DLC-W has been observed to possess variation of resistance against strain sensitivity which was due to deformation of metallic nano-size clusters when a strain is applied enabling its application as a smart material [55]. WC_x is the general chemical composition in which tungsten carbide can exist in DLC doped with tungsten [56]. DLC-W also plays an important role in multi-layer DLC composites by reduction of residual stress and improving adhesion to metallic substrate [57]. It has been noted that DLC-W exhibit tribo-chemical interaction when in contact with lubricants which improves wear resistance and ultra-low friction [16, 51]. The tribological behavior for both room temperature and high temperature (120 and 150) °C were compared for DLC-W [57, 58]. Evaristo et al. [58] observed the presence of tribo-layer on the contact body consisting mainly of W-C, C and W-O acting against the surface of the coatings. However, when dealing with lubricants WS_2 is likely to be formed which serves as tribo-film.

Mo doping of DLC increases the sp^2 carbon content leading to a decrease in the hardness and elastic modulus of DLC-Mo coatings compared with DLC coatings [59]. MoC nanoparticles embedded in the cross-linked amorphous carbon matrix was responsible for the reduction of loss in hardness and elastic modulus [59]. An increase in Mo content increases the surface roughness and decreases the residual stress of the DLC coatings [60]. 3.8% at Mo was reported as the optimum amount for improved mechanical properties [49, 61]. The tribological and mechanical properties of tungsten or molybdenum containing DLC coatings have high hardness, high elasticity and low surface energy, which causes a high hardness to elasticity ratio (H/E) and a low surface energy to hardness ratio (S/H) [52]. Consequently, leading to better adhesion properties, high wear resistance, low friction coefficient and low residual stress of the DLC-W and DLC-Mo [53, 62, 63]. Tribological properties of DLC coatings sometimes maybe affected by adhesion promoter (interlayer), substrate roughness, hydrogen incorporation and coating deposition parameters such as bias voltage (750v), etching, current (1.5A), precursor gas (C2H2/CH4/SiH4/Ar), pressure (2.3 mTorr), time (2 hours) and substrate temperature (200°C) [2]. Service condition such as temperature, sliding speed, load, relative humidity, counter surface and lubrication affects the tribology of DLC-coated parts. As the temperature in the combustion engine increases above 200°C the DLC coatings begin to undergo transformational changes from sp^3 carbon bond to sp^2 carbon bond, making the DLC coating softer to resist hardness and wear, therefore increasing the COF. Although there has been success in the application of DLC coatings for mass production of engine components [48].

3. Present and future DLC coatings

Generally, the DLCs coatings thickness is very important to aid in resistance against wear, which further affects the coating life span. All coatings will eventually wear out due to their exposure to harsh working conditions. Erdemir [64] mentioned that the next generation of hard coatings that can generate their own tribo-films (catalytic coating) in a self-healing or continuous manner will be a big step. If engine blocks can be made from lighter materials the efficiency can be improved, coatings can play a major role in enabling such to work out. The ability to self-generate a DLC film *in situ* during engine operation is possible [64].

3.1 Nanoparticles incorporated DLC coatings

Metal ion can be incorporated into DLC coatings in a controlled manner for relevant laboratory experimental study and industrial use [65]. This also means that metallic nanoparticles can also be incorporated into DLC films, uniform dispersion of nanoparticles into the DLC film is of significant importance [7, 66]. Hexane and isopropanol solution have recently reported to be used to dissolve nanoparticles for DLC incorporation [8]. Using this technique, it is possible to achieve various types of DLCs incorporated with different nanoparticles for different applications ranging from biomedical, electronics, mechanical and automobile engine. TiO_2 nanoparticles incorporated DLC has been reported to increase bacteria activity interaction on DLC surfaces. While, Ag nanoparticles incorporated DLC is known for the increase in wear resistance it provides for DLC [8]. Tungsten trioxide nanocrystalline-containing DLC (WO_3/DLC) has been deposited using one-step electrodeposition technique at atmospheric pressure, which was fabricated for electrical and wetting properties application. The technique makes use of an electrolytic system, a negative electrode (silicon wafer substrate) and positive electrode (platinum plate). The distance between the two electrodes was about 8 mm. Analytical pure methanol (99.5%) was used as carbon source and the concentration of tungsten carbonyl which used as an incorporated reagent was 0.2 mg/ml. The films deposition was carried out under an applied voltage of 1200 V and ambient temperature of 55 ± 2°C for 8 h [11]. Recently, tungsten trioxide (WO_3) and molybdenum trioxide (MoO_3) nanoparticles (DLC nanocomposite coatings) has been incorporated into DLC by first uniformly dispersing these nanoparticles into isopropanol solution and incorporated into DLC coatings deposited on a tappet valve (metallic substrate) deposition parameters for DLC-MoO_3 and DLC-WO_3 coatings is shown in **Table 1** [12, 13]. This tungsten trioxide (WO_3) and molybdenum trioxide (MoO_3) nanoparticles incorporated DLC coatings (DLC-WO_3/DLC-MoO_3) has showed improved adhesion and tribological properties having a COF of 0.075 and 0.070 for DLC-WO_3/DLC-MoO_3 respectively [12, 13]. These coatings with its excellence adhesion and tribological properties is a novel fabricated diamond-like carbon coatings incorporated with nanoparticles using PECVD for automobile applications. The schematic is represented in **Figure 8** below.

Precursor	Pressure (Torr)	Time (min)	Flow rates (sccm)	Bias (-V)
Ar	2.3×10^{-3}	30	8	600
Ar + SiH_4	4.1×10^{-3}	20	3.5	700
Ar + C_2H_2	3.3×10^{-3}	5	10	700
Ar + C_2H_2 + MoO_3/WO_3 nanoparticles	3.6×10^{-3}	55	10	700

Table 1.
Processing parameters for deposition of DLC-MoO3 and DLC-WO3.

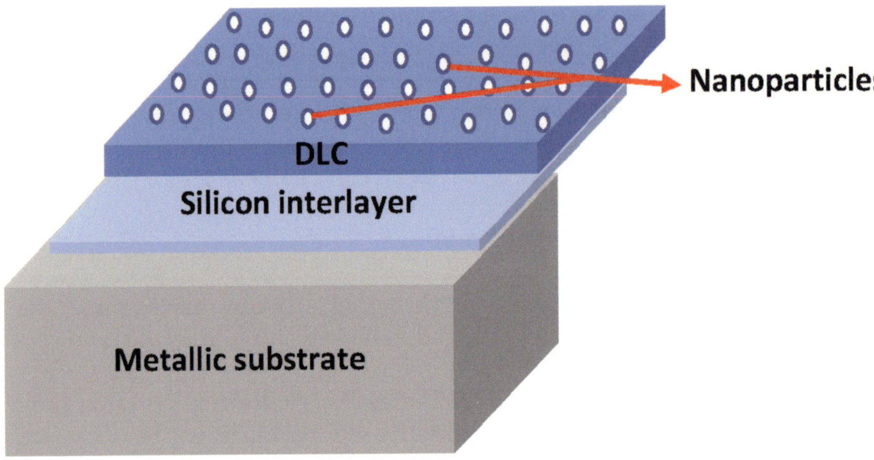

Figure 8.
Schematic of DLC incorporated nanoparticles (DLC nanocomposite coatings).

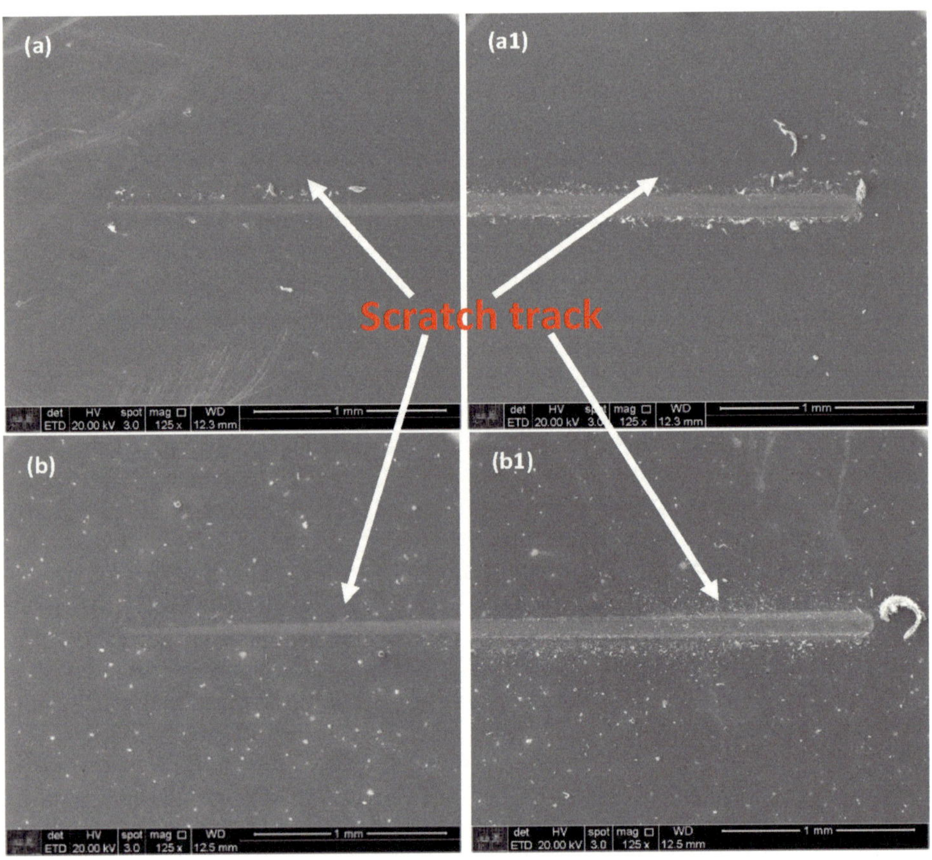

Figure 9.
SEM images of scratch track (a)-(a1) DLC-MoO_3 (b)-(b1) DLC-WO_3.

Nanoparticles (WO_3 or MoO_3) with a higher specific surface would adsorb great amount of cation ions. Under the effect of high electric field, the migration of the cation ions toward to the cathodic substrate with the abundance of electron occurred, resulting in the growth of nc-WO_3 or nc-MoO_3 doped nanocomposite film. **Figure 9** shows the scratch track of DLC-MoO_3 and DLC-WO_3 coatings.

$mCH_3^+ + n(WO_3/MoO_3) + me^- \rightarrow$ nc- (WO_3/MoO_3) doped DLC nanocomposite film.

3.2 Self-healing DLC

The coating with the ability of self-healing itself could be the solution for extending the life of engine components and subsystems. Researchers at the Argonne National Laboratory have developed a self-renewing hard and slick coating for metal parts that has the potential to revolutionize friction and wear protection [64]. The nanocomposite coating uses metal alloys including copper, nickel, palladium and platinum, which are catalytically active at the temperatures at which coatings are used in engines [64]. Structurally, the tribo-film formed during the use of the coatings is similar to diamond-like carbon, which already provides lower friction than industry-standard tribo-film. The DLC coatings interaction with oil molecules makes the DLC film to adhere to metallic substrate. However, in this new technology, when the tribo-film is worn away the catalyst is re-exposed to the lubricant, kick-starting catalysis and developing new layers of tribo-film. Ali Erdemir [64] mention that it might enable additives engine lubricants to provide not only the fluid film, but also the solid boundary film lubrication in a self-healing manner.

4. Conclusion

In this chapter the need for DLC coatings for automobile engine applications has been discussed, the importance of using PECVD technique to deposit DLC coatings has been discussed, together with their mechanical and tribological properties. The thermal stability of DLC coatings was also discussed and with improvement with non-metallic/metallic doping of DLC coatings. Discussion on the recent incorporation of Ag, TiO_2, WO_3 and MoO_3 nanoparticles into DLC. The novel fabrication of diamond-like carbon coatings incorporated nanoparticles (WO_3/MoO_3) using PECVD for automobile applications has shown an improvement in the adhesion properties of the DLC coatings, which will in turn improve the wear resistance of the DLC-WO_3/DLC-MoO_3 coatings.

Acknowledgements

The authors are thankful to the Petroleum Technology Development Fund (PTDF), Nigeria and Conselho Nacional de Desenvolvi-mento Científico e Tecnologico - Brasil (CNPq) process 141991/2019-4 and 315861/2018-5. In addition, the authors are grateful to Instituto Nacional de Pesquisas Espaciais (INPE), Sao Jose dos Campus, Brazil for the DLC film deposition.

Author details

Funsho Olaitan Kolawole[1,2]*, Shola Kolade Kolawole[3,4], Luis Bernardo Varela[1,5], Adebayo Felix Owa[2], Marco Antonio Ramirez[6] and André Paulo Tschiptschin[1]

1 Department of Metallurgical and Materials Engineering, University of São Paulo, Sao Paulo, SP, Brazil

2 Department of Materials and Metallurgical Engineering, Federal University, Oye-Ekiti, Nigeria

3 National Agency for Science and Engineering Infrastructure, Abuja, Nigeria

4 Materials Science and Engineering, African University of Science and Technology, Abuja, Nigeria

5 Vale Institute of Technology, Ouro Preto, MG, Brazil

6 Instituto de Pesquisa e Desenvolvimento, Universidade do Vale do Paraiba, São Jose dos Campos, Brazil

*Address all correspondence to: funsho.kolawole@usp.br

References

[1] Holmberg K, Andersson P, Erdemir A. Tribology International Global energy consumption due to friction in passenger cars. Tribiology International. 2012;47:221-234.

[2] Al Mahmud KAH, Kalam MA, Masjuki HH, Mobarak HM, Zulkifli NWM. An updated overview of diamond-like carbon coating in tribology. Critical Review in Solid State and Materials Science. 2005;40(2):90-118.

[3] Hamedani Y, Macha P, Bunning TJ, Naik RR, Vasudev MC. Plasma-Enhanced Chemical Vapor Deposition: Where we are and the Outlook for the Future in Chemical Vapor Deposition - Recent Advances and Applications in Optical, Solar Cells and Solid State Devices. London: IntechOpen. 2016;247 – 280.

[4] Lawes SDA, Fitzpatrick ME, Hainsworth SV. Evaluation of the tribological properties of DLC for engine applications. Journal of physics D: Applied physics. 2007;40:5427-5437.

[5] Bewilogua K, Hofmann D. History of diamond-like carbon films – from first experiments to worldwide applications. Surface & Coatings Technology. 2014;242:214-225.

[6] Dearnley PA, Neville A, Turner S, Scheibe HJ, Tietema R, Tap R, Stuber M, Hovsepian P, Layyous A, Stenbom B. Coatings tribology drivers for high density plasma technologies. Surface Engineering. 2010;26(1-2):80-96.

[7] Hussain S, Roy RK, Pal AK. Incorporation of silver nanoparticles in DLC matrix and surface plasmon resonance effect. Materials Chemistry and Physics. 2006;99:375- 381.

[8] Marciano FR, Lima-Oliveira DA, Da-Silva NS, Diniz AV, Corat EJ, Trava-Airoldi VJ. Antibacterial activity of DLC films containing TiO_2 nanoparticles. Journal of Colloid and Interface Science. 2009;340:87-92.

[9] Wana S, Wanga L, Xue Q. Super-hydrophilic properties of TiO_2–DLC nanocomposite films fabricated by the simple electrochemical process. Applied Surface Science. 2011;257:10000-10004.

[10] Ban M, Hasegawa N. Deposition of diamond-like carbon thin films containing photocatalytic titanium dioxide nanoparticles. Diamond & Related Materials, 2012; 25:92-97.

[11] Huang D, Wan S, Wang L, Xue Q. One-step electrodeposition of amorphous carbon films containing WO_3 with high conductivity and good wettability. Surface and Interface Analysis. 2010;43(7):1-18.

[12] Kolawole FO, Varela LB, Kolawole SK, Ramirez MA, Tschiptschin AP. Deposition and characterization of tungsten oxide (WO_3) nanoparticles incorporated diamond-like carbon coatings using pulsed-DC PECVD. Materials Letters. 2021;282(128645):1 - 4.

[13] Kolawole FO, Ramirez MA, Kolawole SK, Varela LB, Tschiptschin AP. Deposition and characterization of molybdenum oxide (MoO_3) nanoparticles incorporated diamond-like carbon coatings using pulsed-DC PECVD. Materials Letters. 2020;278(128420):1 - 4.

[14] Moriguchi H, Ohara H, Tsujioka M. History and Application of Diamond-like Carbon Manufacturing Processes. SEI Technical Review. 2016;82:52-58.

[15] AwadeshKM, NanadadulalD, PrajitG, Utpal G, Sukhendu J, Sayan D, Kaustav G, Garfield R, Samir KL, Someswar D. Deposition and Characterization of

Diamond-like-Nanocomposite grown by Plasma Enhanced Chemical Vapor Deposition over Different Substrate Materials. Bulletin of Materials Science. 2013:36(2):193-202.

[16] Zahid R, Masjuki HH, Varman M, Mufti RA, Kalam MA, Gulzar M. Effect of lubricant formulations on the tribological performance of self-mated doped DLC contacts: A review. Tribology Letters. 2015;58(32):1-28.

[17] Yasuda Y, Kano M, Mabuchi Y, Abou S. Research on diamond-like carbon coatings for low friction valve lifters. SAE Technical Paper Series. 2016;1-9.

[18] Louda P. Applications of thin coatings in automotive industry. Journal of achievements in materials and manufacturing engineering. 2007;24(1):51-56.

[19] Koidl P, Wagner C, Dischler B, Wagner J, Ramsteiner M. Plasma deposition, properties and structure of amorphous hydrogenated carbon films. Materials Science Forum. 1990:52:41-70.

[20] Silva GF. Estudos dos Mecanismos que Promovem a Aderência de DLC na Superfície de Cobre. Dissertação (Mestrado em ETE/Ciência e Tecnologia dos Materiais e Sensores). Instituto Nacional de Pesquisas Espaciais (INPE), São José dos Campos. 2013:101.

[21] Robertson J. Diamond-like amorphous carbon. Materials Science and Engineering. 2002; R37 (4-6):129-281.

[22] Trava-Airoldi V, Bonetti L, Capote G, Santos LV, Corat EA. Comparison of DLC film properties obtained by r.f. PACVD, IBAD, and enhanced pulsed-DC PADVC. Surface and Coatings Technology. 2007;202(3):549-554.

[23] Capote G, Bonetti LF, Trava-Airoldi VJ, Santos LV, Corat EJ. Deposition of adherent DLC films using a low-cost enhanced pulsed-DC PECVD method. Brazilian Journal of Vacuum Applications. 2006;25(4):209-213.

[24] Bonetti LF, Capote G, Santos LV, Corat EJ, Trava-Airoldi, VJ. Adhesion studies of diamond-like carbon films deposited on Ti6Al4V substrate with a silicon interlayer. Thin Solid Films. 2006;515(1):375-379.

[25] Capote GR. Produção e caracterização de filmes finos de carbono amorfo hidrogenado, depositados em plasmas de metano diluídos por gases nobres. Tese (Doutorado em Física) – Pontifícia Universidade Católica do Rio de Janeiro, Rio de Janeiro. 2003;181.

[26] Mobarak HM, Chowdhury M. Tribological performance of hydrogenated amorphous carbon (a-C:H) DLC coating when lubricated with biodegradeable vegetal canola oil. Tribology in Industry. 2014;36(2):163-171.

[27] Matthews A, Eskildsen SS. Engineering applications for diamond-like carbon. Diamond Related Materials. 1994;3(4-6):902-911.

[28] Treutler CPO. Industrial Use of Plasma-Deposited Coatings for Components of Automotive Fuel Injection Systems. Surface & Coatings Technology. 2005;200(5-6):1969-1975.

[29] Ferrari, CA. Diamond-like carbon for magnetic storage disks. Surface and Coatings Technology. 2004;180-181:190-206.

[30] Kim DW, Kim KW. Effects of sliding velocity and normal load on friction and wear characteristics of multi-layered diamond-like carbon (DLC) coating prepared by reactive sputtering. Wear. 2013;297(1-2):722-730.

[31] Goncalves PAR. Triboquímica Analítica: Estudo da Dependência dos Parâmetros Tribológicos com a Modificação Superficial em Filmes a Base de Carbono. Tese (Doutorado em ETE/ Ciência e Tecnologia de Materiais e Sensores). Instituto Nacional de Pesquisas Espaciais (INPE), São José dos Campos, 2012;196. Disponível em: <http://urlib.net/8JMKD3MGP7W/3CA9SRH>

[32] Choy KL, Felix E. Functionally graded diamond-like carbon coatings on metallic substrates. Materials Science and Engineering: A. 2000;2789(1-2):162-169.

[33] Silva WM, Modificação da superfície do aço para melhorias na aderência de filmes de DLC. Tese (Doutorado em ETE/ Ciência e Tecnologia de Materiais e Sensores) – Instituto Nacional de Pesquisas Espaciais (INPE), São José dos Campos. 2011. 96. Disponível em: <http://urlib.net/8JMKD3MGP7W/396CU8E>.

[34] Wang P, Wang X, Xu T, Liu W, Zhang J. Comparing internal stress in diamond-like carbon films with different structure. Thin Solid Films. 2007;**515**(17):6899-6903

[35] Ollendorf H, Schneider, DA. Comparative study of adhesion test methods for hard coatings. Surface and Coatings Technology. 1999;113(1-2):86-102.

[36] Alarcon O, Weingaertner W, Roman H, Sousa F, Pereira M, Spiller A, Tridapalli D. Design and development of an apparatus to measure the scratch hardness of ceramic floor tiles. Qualicer. 2004;(2):283-295.

[37] Gahr KHZ. Microstructure and Wear of Materials. Amsterdam: Elsevier Science Publishers B. V. ISBN (0-444-42754-6). Tribology Series. 1987;560.

[38] Ferrari AC, Kleinsorge B, Morrison NA, Hart A. Stress reduction and bond stability during thermal annealing of tetrahedral amorphous carbon. Journal Applied Physics. 1999;85(10):7191-7719.

[39] Kalish R, Lifshitz Y, Nugent K, Prawer S. Thermal stability and relaxation in diamond-like-carbon. A Raman study of films with different sp^3sp^3 fractions (ta-C to a-C). Applied Physics Letter. 1999;74(20):2936-2938.

[40] Vetter J. 60 years of DLC coatings: Historical highlights and technical review of cathodic arc processes to synthesize various DLC types, and their evolution for industrial applications. Surface & Coatings Technology. 2014;257:213-240.

[41] Strel'nitskij VE, Vakula SI, Matjushenko NN. Ukrainian Journal of Physics. 1977;22:431-433.

[42] Tay BK, Sheeja D, Lau SP, Shi X, Seet BC, Yeo YC. Time and temperature-dependent changes in the structural properties of tetrahedral amorphous carbon films. Surface & Coatings Technology. 2000;130(2-3):248-251.

[43] Anders S, Ager III JW, Pharr GM, Tsui TY, Brown IG. Heat treatment of cathodic arc deposited amorphous hard carbon films. Thin Solid Films. 1997;308-309:186-190.

[44] Lee S, Yeo CD. Microwear mechanism of head carbon film during head disk interface sliding contact. Tribology International. 2012;45:30-37.

[45] VDI 2840, Carbon films - Basic knowledge, film types and properties, June 1, 2012.

[46] Pu JC, Wang SF, Sung JC. High-temperature oxidation behaviors of CVD diamond films. Applied Surface Science. 2009;256(3):668-673.

[47] Balden M, Klages KU, Jacob W, Roth J. Oxidation erosion of graphite in

air between 600 and 1000 K. Journal of Nuclear Materials. 2005;341:31 – 44.

[48] Kano M. Overview of DLC coated engine components. In S.C. Cha, A. Erdemir, (Eds.), Coating technology for vehicle applications. Cham, Switzerland, Springer International Publishing AG. 2015;39-40.

[49] Fu KY, Mei YF, Fu MY, Liu XY, Chu PK. Thermal stability of metal-doped diamond-like carbon fabricated by dual plasma deposition. Diamond & Related Materials. 2005;14:1489-1493.

[50] Cemin F, Bim LT, Menezes CM, Maia da Costa MEH, Baumvol IJR, Alvarez F, Figueroa CA. The influence of different silicon adhesion interlayers on the tribological behavior of DLC thin films deposited on steel by EC-PECVD. Surface & Coatings Technology. 2015;283:115-121.

[51] Yang L, Neville A, Brown A, Ransom P, Morina A. Friction reduction mechanisms in boundary lubricated W-doped DLC coatings. Tribology International. 2014;70:26-33.

[52] Donnet C. Recent progress on the tribology of doped diamond-like and carbon alloy coatings: a review. Surface and coatings technology. 1998;100-101:180-186.

[53] Havsepian P, Mandal P, Ehiasarian A, Safran G, Tietema R, Doerwald D. Friction and wear behaviour of Mo – W doped carbon-based coating during boundary lubricated sliding. Applied surface science. 2016;366:260-274.

[54] Bhowmick S, Banerji A, Alpas AT. Tribological Behaviour of W-DLC against an Aluminium Alloy Subjected to Lubricated Sliding. Tribology in Industry. 2015;37(3):277-283.

[55] Takeno T, Miki H, Sugawara T, Hoshi Y, Takagi T. A DLC/W-DLC multilayered structure for strain sensing applications. Diamond & Related Materials. 2008;17:713-716.

[56] Dai M, Wei C, Zhou K, Zhu M, Hou H, Lin S, Tong X. Properties of W/DLC/W-S-C composite films fabricated by magnetron sputtering. Transaction of Nonferrous Metals Society China. 2015;25:3002-3011.

[57] Mutafov P, Lanigan J, Neville A, Cavaleiro A, Polcar T. Surface & Coatings Technology DLC-W coatings tested in combustion engine — Frictional and wear analysis. Surface & Coatings Technology. 2014;260:284-289.

[58] Evaristo M, Fernandes F, Cavaleiro A. Room and High Temperature Tribological Behaviour of W-DLC Coatings Produced by DCMS and Hybrid DCMS-HiPIMS Configuration. Coatings. 2020;10(319):1-15.

[59] Ji L, Li H, Zhao F, Chen J, Zhou H. Microstructure and mechanical properties of Mo/DLC nanocomposite films. Diamond & Related Materials. 2008;17:1949-1954.

[60] Butt MZ, Khaleeq-ur-Rahman M, Ali D, Akmal A, Naseem S. Deposition and characterization of multilayer DLC:Mo thin films grown on silicon substrate by off-axis pulsed laser deposition technique. Applied Surface Science. 2015;331:407-414.

[61] Tang XS, Wang HJ, Feng L, Shao LX, Zou CW. Mo doped DLC nanocomposite coatings with improved mechanical and blood compatibility properties. Applied Surface Science. 2014;311:758-762.

[62] Sanchez-Lopez JC, Fernandez A. Effects on DLC Coatings. In: Donnet C., Erdemir A. (eds) Tribology of Diamond-like carbon films. Springer, Boston, MA. 2008.

[63] Yong Q, Ma G, Wang H, Chen S, Xu B. Influence of tungsten content on microstructure and properties of tungsten-doped graphite-like carbon films. Journal of Materials Research. 2016;31(23):3766-3776.

[64] Donlevy M. Coating Technology. Engine Technology International. accessed 22 June 2019, 42-46. https://www.enginetechnologyinternational.com/online-magazines/in-this-issue-june-2019.html

[65] Schmid G. Large Clusters and Colloids. Metals in the Embryonic State. Chemical Reviews. 1992;92:1709-1727.

[66] Schiffmann KI, Fryda M, Goerigk G, Lauer R, Hinze P, Bulack A. Sizes and distances of metal clusters in Au-, Pt-, W- and Fe-containing diamond-like carbon hard coatings: a comparative study by small angle X-ray scattering, wide angle X-ray diffraction, transmission electron microscopy and scanning tunnelling microscopy. Thin Solid Films. 1999;347:60-71.